Biotic Stress Management of Crop Plants using Nanomaterials

This book summarizes nanotechnology-based agricultural research for crop productivity and the management of various plant pathogens. It deals with the application of nano-molecules for quick, cost-effective, and precise plant disease diagnostic procedures, plant pests and disease management, nano-pesticides, and nano-diagnostics. Further, it explains nanomaterials for biotic stress management, with an insight into the synthesis and modification of nanomaterials and their potential applications in different domains for disease management.

Features include:

- Compilation of current research on nanomaterials as well as their versatile applications in plant biotic stress management
- Description of the role of nanomaterials as enzyme-mimicking nanoparticles, nano-pesticides, nano-fertilizers, and nanomaterials
- Review of day-to-day problems related to crop plants, their diagnostics, and stress management
- Exploration of trends in nanomaterial utility in diagnostics, enzyme-mimicking, and crop protection, and their possible role in plant disease management
- Study of pertinent nanomaterials including synthetic strategies, properties, chemistry, and applications

This book is aimed at researchers and graduate students in plant pathology, genetic engineering, environmental science, botany, bioengineering, and nanotechnology.

Advances in Bionanotechnology

Series Editors: Ravindra Pratap Singh, *Department of Biotechnology, Indira Gandhi National Tribal University, Anuppur, Madhya Pradesh, India,* **Jay Singh**, *Department of Chemistry, Institute of Science, Banaras Hindu University, Varanasi, Uttar Pradesh, India* and **Charles Oluwaseun Adetunji**, *Department of Microbiology, Edo State University Uzairue, Iyamho, Edo State, Nigeria*

Bionanotechnology is a multi-disciplinary field that shows immense applicability in different domains, namely chemistry, physics, material sciences, biomedical, agriculture, environment, robotics, aeronautics, energy, electronics and so forth. This book series will explore the enormous utility of bionanotechnology for biomedical, agricultural, environmental, food technology, space industry, and many other fields. It aims to highlight all the spheres of bionanotechnological applications and its safety and regulations for using biogenic nanomaterials that are a key focus of the researchers globally.

Bionanotechnology Towards Sustainable Management of Environmental Pollution
Edited by Naveen Dwivedi and Shubha Dwivedi

Natural Products and Nano-formulations in Cancer Chemoprevention
Edited by Shiv Kumar Dubey

Bionanotechnology towards Green Energy
Innovative and Sustainable Approach
Edited by Shubha Dwivedi and Naveen Dwivedi

Biotic Stress Management of Crop Plants using Nanomaterials
Edited by Krishna Kant Mishra and Santosh Kumar

For more information about this series, please visit: www.routledge.com/Advances-in-Bionanotechnology/book-series/CRCBIONAN

Biotic Stress Management of Crop Plants using Nanomaterials

Edited by
Krishna Kant Mishra and Santosh Kumar

CRC Press
Taylor & Francis Group
Boca Raton London New York

CRC Press is an imprint of the
Taylor & Francis Group, an **informa** business

First edition published 2023
by CRC Press
6000 Broken Sound Parkway NW, Suite 300, Boca Raton, FL 33487–2742

and by CRC Press
4 Park Square, Milton Park, Abingdon, Oxon, OX14 4RN

CRC Press is an imprint of Taylor & Francis Group, LLC

ISBN: 9781032344317 (hbk)
ISBN: 9781032344324 (pbk)
ISBN: 9781003322122 (ebk)

DOI: 10.1201/9781003322122

Typeset in Times
by Newgen Publishing UK

Contents

Preface

The current technical era demands more pronounced techniques to achieve increased demand for sustainable agriculture. In the current scenario, global agriculture is facing numerous unprecedented challenges. Therefore, it becomes necessary to overcome the barriers to feeding the world and focus on addressing various challenges. In order to achieve food security, nanotechnology is a handy tool for boosting crop production and has contributed to the agro-technological revolution. Nanotechnology is being used widely in all disciplines of science and technology, including plant sciences. Many methods and routes of nanomaterial synthesis have been reported from various materials. Nanotechnology may help improve crop production by maintaining plant health. Various nano-molecule-based formulations have been widely investigated for plant health management. Nano-molecules offer a wider specific surface area to fertilizers and pesticides. As unique carriers of agrochemicals, they facilitate the site-targeted controlled delivery of nutrients with enhanced crop protection. However, to increase the resistance of plants without harming the environment, agrochemical release of should be done in a controlled way; for this, nanotechnology has been proved to be a novel tool for achieving an eco-agriculture approach. It plays a protective role for plants against abiotic and biotic stresses.

The purpose of this book is to stimulate further research into the use of nanomaterials in biotic stress management and to foster further interest for researchers, academicians, and scientists worldwide. In this book, current nanotechnology-based agricultural research that could benefit crop productivity and the management of various plant pathogens is addressed. This book also deals in detail with the application of nano-molecules for quick, cost-effective, and precise plant disease diagnostic procedures, applications of nanotechnology in plant pest and disease management, and nano-pesticides. Crops treated with safer nano-pesticides will gain added value since they will be free from pesticidal residues, decay, and putative pathogens harming human health, sustaining the global demand for high product quality.

We hope that this book will serve as an important source of information for researchers, academicians, and scientists to enhance their knowledge in the field of biotic stress management of crop plants using nanomaterials. The exhaustive discussion presented here will help the understanding of the multiple roles of nanoparticles. The information contained in the book will go a long way towards exploiting the role of nanoparticles in crop protection programs.

Krishna Kant Mishra and Santosh Kumar

About the Editors

Krishna Kant Mishra is currently working as Principal Scientist and Incharge Head, Crop Protection Division in Indian Council of Agricultural Research (ICAR)-Vivekananda Parvatiya Krishi Anusandhan Sansthan, Almora, Uttarakhad, India. Dr. Mishra received his B.Sc. (Biology) degree from the University of Allahabad, Allahabad, Uttar Pradesh, India with Gold Medal. He obtained his M.Sc. (Plant Pathology) and Ph.D. in Plant Pathology with minor in Molecular Biology and Genetic Engineering from G.B. Pant University of Agriculture and Technology, Pantnagar, Uttarakhand, India. He worked as Junior Research Officer at the Department of Plant Pathology, G.B. Pant University of Agriculture and Technology from 2006 to 2010. He was given the P.R. Verma Award by the Indian Society of Mycology and Plant Pathology, Udaipur, India in 2006 for his Ph.D. thesis, as well as the Young Scientist Award by Uttarakhand State Science and Technology Congress in 2011. During his career, Dr. Mishra has published more than 50 research papers, two edited books, two authored books, three technical bulletins, more than 30 book chapters, 30 extension leaflets, and more than 50 popular articles. He has attended and participated in many local and international workshops and conferences. He was elected the Zonal President-2021, Mid-Eastern Zone by the Indian Phytopathological Society, New Delhi. He has served as a member of organizing committees and as editorial board member and reviewer for a number of scientific journals.

Santosh Kumar is currently working as Associate Professor at the Department of Chemistry, Harcourt Butler Technical University, Kanpur, India. He has also served as Research Professor in the Department of Organic and Nano System Engineering, Division of Chemical Engineering, Konkuk University, Seoul, South Korea. Dr. Kumar received his B.Sc. and M.Sc. in Chemistry and D.Phil. in Organic Chemistry from the University of Allahabad, India. He worked as a Research Associate (Council of Scientific and Industrial Research: CSIR) at the Department of Chemistry, Motilal Nehru National Institute of Technology, Allahabad, India. He was awarded a Chinese Academy of Sciences–The World Academy of Sciences (CAS-TWAS) postdoctoral fellowship in 2008 and worked in Changchun Institute of Applied Chemistry, Chinese Academy of Sciences, Changchun, China. Later he joined as a Konkuk University Brain Pool Professor and researcher at the Konkuk University, Seoul, South Korea (2010–2013). He was awarded a Fundação para a Ciência a Tecnologia (FCT) Postdoctoral Fellowship in 2012 and worked at the Chemistry Centre, University of Coimbra, Coimbra, Portugal from 2013 to 2017. During his career, Dr. Kumar has published more than 82 research papers, edited one book and seven book chapters, and attended and participated in local and international workshops and conferences. He has served as a member of organizing committees and an editorial board member and reviewer for a number of scientific journals. He is Associate Editor of the *Asian Chitin Journal* and is Editor-in-Chief of *Journal of Biological Engineering Research and Review*. Current research interests involve source apportionment of chemical modification, optical and biological properties of chitosan biopolymer for biomedical applications, hydrogel, aerogel, nanomaterials, drug and gene delivery, and environmental chemistry.

Contributors

Azhar Abbas, Department of Chemistry, University of Sargodha, Sargodha 40100, Pakistan

Sougata Bhattacharjee, ICAR-Indian Agricultural Research Institute, Hazaribagh, India

Rakesh Bhowmick, Crop Improvement Division, ICAR-Vivekananda Parvatiya Krishi Anusandhan Sansthan, Almora, India

Mehrdad Chaichi, Hamedan Agriculture and Natural Resources, Research and Education Center, Agricultural Research, Education and Extension Organization, Hamedan, Iran

Rahul Dev, Crop Improvement Division, ICAR-Vivekananda Parvatiya Krishi Anusandhan Sansthan, Almora-263601, India

Bipratip Dutta, ICAR-National Institute of Plant Biotechnology, New Delhi, India

Mohadeseh Hassanisaadi, Department of Plant Protection, Faculty of Agriculture, Shahid Bahonar University of Kerman, Kerman, Iran

Moslem Heydari, Department of Plant Production and Genetics, University of Zanjan, Zanjan, Iran

Muhammad Ikram, Department of Botany, PMAS Arid Agriculture University, Rawalpindi, Punjab 46300, Pakistan

Jeevan Bettanayaka, Crop Protection Division, ICAR-Vivekananda Parvatiya Krishi Anusandhan Sansthan, Almora, Uttarakhand, India

Lakshmi Kant, Crop Improvement Division, ICAR-Vivekananda Parvatiya Krishi Anusandhan Sansthan, Almora-263601, India

Hamed Hassanzadeh Khankahdani, Horticulture Crops Research Department, Hormozgan Agricultural and Natural Resources Research and Education Center, Agricultural Research, Education and Extension Organization (AREEO), Bandar Abbas, Iran

Amit Kumar, Crop Production Division, ICAR-Vivekananda Parvatiya Krishi Anusandhan Sansthan, Almora, India

Santosh Kumar, Department of Chemistry, Harcourt Butler Technical University, Kanpur-208002, U.P., India

Zia-Ur-Rahman Mashwani, Department of Botany, PMAS Arid Agriculture University, Rawalpindi, Punjab 46300, Pakistan

Soheila Mirzaei, Department of Plant Protection, Faculty of Agriculture, Bu-Ali Sina University, Hamedan, Iran

Krishna Kant Mishra, Crop Protection Division, ICAR-Vivekananda Parvatiya Krishi Anusandhan Sansthan, Almora, Uttarakhand, India

Azza H. Mohamed, Agricultural Chemistry Department, Faculty of Agriculture, Mansoura University, Mansoura 35516, Egypt; Citrus Research and Education Center, University of Florida, IFAS, Lake Alfred, FL 33850, United States of America

Tilak Mondal, Crop Production Division, ICAR- Vivekananda Parvatiya Krishi Anusandhan Sansthan, Almora, India

Ahmad A. Omar, Citrus Research and Education Center, University of Florida, IFAS, Lake Alfred, FL 33850, United States of America; Biochemistry Department, Faculty of Agriculture, Zagazig University, Zagazig 44519, Egypt.

R S Pal, Crop Improvement Division, ICAR-Vivekananda Parvatiya Krishi Anusandhan Sansthan, Almora-263601, India

Manoj Parihar, Crop Production Division, ICAR-Vivekananda Parvatiya Krishi Anusandhan Sansthan, Almora, India

Amit Umesh Paschapur, Crop Protection Division, ICAR-Vivekananda Parvatiya Krishi Anusandhan Sansthan, Almora, Uttarakhand, India

Baswaraj Raigond, ICAR-Indian Institute of Millets Research, Research Station, Solapur, Maharashtra, India

Naveed Iqbal Raja, Department of Botany, PMAS Arid Agriculture University, Rawalpindi, Punjab 46300, Pakistan

Rashmi Rameshwari, Department of Biotechnology, Manav Rachna International Institute of Research and Studies, Faridabad, India

Seyedeh-Somayyeh Shafiei-Masouleh, Department of Genetics and Breeding, Ornamental Plants Research Center (OPRC), Horticultural Sciences Research Institute (HSRI), Agricultural Research, Education and Extension Organization (AREEO), Mahallat, Iran

Devender Sharma, Crop Improvement Division, ICAR-Vivekananda Parvatiya Krishi Anusandhan Sansthan, Almora-263601, India

Ayesha Siddiqa, Department of Botany, PMAS Arid Agriculture University, Rawalpindi, Punjab 46300, Pakistan

Ashish Kumar Singh, Crop Protection Division, ICAR-Vivekananda Parvatiya Krishi Anusandhan Sansthan, Almora, Uttarakhand, India

Devendra Kumar Verma, Department of Chemistry, Sri Venkateswara College, University of Delhi (South Campus), Delhi, India

Gaurav Verma, Crop Protection Division, ICAR-Vivekananda Parvatiya Krishi Anusandhan Sansthan, Almora, Uttarakhand, India

Seyed Morteza Zahedi, Department of Horticultural Science, Faculty of Agriculture, University of Maragheh, Maragheh, Iran

Efat Zohra, Department of Botany, PMAS Arid Agriculture University, Rawalpindi, Punjab 46300, Pakistan

1 Overview of Nanomaterials and their Synthesis

Ayesha Siddiqa, Muhammad Ikram, Efat Zohra,
Naveed Iqbal Raja, Zia-Ur-Rahman Mashwani,
Azza H. Mohamed, Seyed Morteza Zahedi,
Azhar Abbas, and Ahmad A. Omar

CONTENTS

DOI: 10.1201/9781003322122-1

1.1 INTRODUCTION

Nanomaterials are the foundation of nanotechnology and nanoscience. Nanotechnology is a major multidisciplinary area of research that has been developing globally over the past few decades. Nanotechnology has the potential to revolutionize the procedure through which different products are created and increase their range and nature of function. Moreover, this science has a significant economic impact that will revolutionize the future. Nanoparticles (NPs) are defined as a set of substances where at least one dimension is less than approximately 100 nm. A nanometer is one-millionth of a millimeter, which gives it unique electrical, optical, and magnetic properties [1]. These developing properties have the potential for significant impact in numerous fields, such as medicine, agriculture, and electronics. Some NPs occur in nature but most are engineered according to a particular interest for their use in different commercial products. NPs are widely used in the manufacture of different products such as sports goods, cosmetics, sun-block creams, stain-resistant clothes, and electronics, and in the field of medicine for drug delivery, diagnosis, and treatment of disease. Because of their small size, NPs have a greater surface area-to-volume ratio, which ultimately enhances their chemical reactivity and strength [2].

1.2 HISTORY OF NANOMATERIALS

Any material having a size in the range of 1–100 nm is termed nanomaterial. The history of nanomaterials began with a big bang (a process that led to the formation of the earth), as it is believed that nanomaterial originates from meteorites [3]. Naturally occurring nanomaterials were invented in prehistoric times, when human beings used fires and the smoke of such fires contained various types of nanomaterial. Forest fire products, the weathering process of different types of rock, radioactive decay, and volcanic ash are among the natural sources of nanomaterial.

Scientific studies of nanomaterials began much later. One of the first scientific syntheses of colloidal gold NPs was done by Michael Faraday in 1857. In the earliest time in Egypt, a black pigment which had high opacity and stability was produced from the soot of oil lamps for writing on paper or papyrus [4]. At that time, the Egyptians were unaware of the fact that soot contains carbon nanomaterials. They also manufactured PdS_2 NPs with a diameter of approximately 5 nm as a component of hair colorant using different synthetic chemical processes. Metallic nano-powders were synthesized in the 1960s and 1970s for use in magnetic recorder tapes. Our ancestors unknowingly used nanomaterials such as carbon for cave painting. In the earliest times, Egyptians produced synthetic pigments in the 5-nm thick nano-sheets by mixing nano-sized particles of glass and quartz [5]. Egyptian blue pigment (nanomaterial) was widely used for decorative purposes in countries that were part of the Roman empire. First, metallic NPs were produced by Egyptians using different chemical methods in the 13th and 14th centuries. Besides these Celtic red enamels which contain Cu, NPs along with cuprous oxide were synthesized between 400 and 100 BC. Mayan blue is a pigment that is extremely resistant to corrosion; it was first synthesized in 800 AD [6]. It is a composite nanomaterial containing a mixture of indigo dye which is chemically linked with clay nano-pores to obtain a stable dye. Carbon nanotubes were found in the microstructure of Wootz steel that was first produced in India in 900 AD. For many centuries lusterware was famous and widely used in Muslim countries, as well as in some parts of Europe; later on, elemental analysis showed the presence of Ag, Cu, and other NPs. Transmission electron microscopic (TEM) studies revealed the presence of two thin layers of Ag NPs towards the outside and two thick layers towards the inside.

With the development of different characterization techniques, it has become possible to study the characteristics of nanomaterials, to assess their origin and their impact on the earth's environment in a better way. Modern nanotechnology began in the 1960s, although nanomaterials had been made and used in ancient civilizations. In the modern age, various nanomaterials have been developed to enhance the properties of bulk material such as strength, toughness, and lightness, and also to add new features to the material [7]. In addition to all these advantages, NPs have a broad range of applications in different fields due to their small size and specific shape. Nano-phase cluster material is synthesized by preparing small mixtures which are then fused to prepare bulk material with outstanding features. These are used in building materials as safety sensing components and for other biomedical purposes such as biosensors and drug delivery, as vehicles to deliver a drug to the target site. Moreover, with the use of nanotechnology, soon it will be possible for the military to create a sensor system that would detect any drug or biological agent with high sensitivity and low detection limits. In 2003, NP-based new metallic or non-metallic coating paint was produced to enhance car resistance to scratches. In 2014 more than 800 nanotechnology-based products were introduced and used in more than 20 countries. It is estimated that the nanomaterial market will expand tremendously during the period 2021–2027 due to increasing use in industrial sectors [4].

1.3   PROPERTIES OF NANOMATERIALS

Nanomaterials possess unique features and their properties differ significantly from bulk materials. This is mainly due to their nano size, which renders high surface energy, large surface area, and spatial confinement. These properties usually do not occur in bulk material. The size and surface area of any material determine the way it interacts with any biological system. The decrease in the size of the material increases the surface area and enhances its ability to react with any surface comes in contact with. It has been observed that the efficiency of various biological systems depends on the size of the material [8]. Cytotoxicity of NPs has been evaluated by researchers using in vivo models; the researchers found that cytotoxicity is due to their small size. As a result they can easily enter any biological system and destroy the structure of various macromolecules. Nanomaterials cause a dramatic change in the properties of bulk material. For example, the use of NPs in chemical sensors enhanced their sensitivity and sensor selectivity. The use of NPs modified the energy band structure and charge carrier density of bulk material, which in turn altered the electronic and optical properties of large materials. The outstanding mechanical properties of nanomaterials are due to their volume, surface, and quantum effect [9]. The addition of NPs to any material causes the formation of an intra-granular structure, improving the mechanical properties of the material. One useful aspect of nanomaterials is their optical properties. Different types of NPs exhibit a strong ultraviolet (UV)-visible extinction band that is not found in a spectrum of large-size material.

There is increasing interest in exploring the magnetic properties of NPs. Previous work revealed that NPs perform best when their size is in the range of 10–20 nm. At this level, the magnetic properties of NPs dominate and make them priceless for use in different fields [10]. These mechanical properties of NPs compel researchers to look for their applications in the field, such as nanofabrication and nano-manufacturing. Mechanical properties such as hardness, stress, strain, and friction can be studied to find the exact mechanism of action. By controlling the mechanical properties and their interaction with different surfaces, the surface quality of large materials can be enhanced. Electrical properties of nanomaterials exhibit electrical conductance through nanotubes and nanowires, and give information about the electrical conductivity of a nanocomposite. Electrical conductance can be measured by the mechanical thinning of wires and measuring the current passing through them at applied voltage [11]. With the decrease in the diameter of a tube, the number of free electrons reduces, conductivity becomes less and as a result electric conductance decreases. It has been observed that NPs (in fluid form) exhibit more thermal conductivity than a fluid with large-size material. Nano-fluids are usually produced by dispersing nano-sized solid particles in a liquid medium such as ethylene, glycol, or oil. Heat transfer takes place through the surface and NPs have a large surface area so they show superior thermal properties compared to fluid with bulky material [12].

1.4 CLASSIFICATION OF NANOMATERIALS

Nanomaterials have been classified based on various criteria, such as morphology, dimension, and chemical composition. Size is also an important attribute for the classification of nanomaterials. Classes of nanomaterials include: (1) carbon-based nanomaterials; (2) metal-based nanomaterials; (3) semiconductor nanomaterials; and (4) nanocomposite.

1.4.1 CARBON-BASED NANOMATERIALS

Carbon-based nanomaterials contain carbon as the main component and play a key role in various fields. They occur in different forms, such as fullerene, carbon-based nanotubes, and graphene. Carbon nanomaterials have a vast range of applications on different technical grounds, such as nano-electronics, gas storage, production of conductive plastics, antifouling paints, and batteries with increased durability [13]. Fullerenes are allotropic modifications of carbon atoms that consist of a group of atomic carbons in the form of a cluster with a carbon atom on a spherical surface.

In fullerene carbon atoms with sp^2 hybridization are linked with each other through a covalent bond. Carbon atoms are located at apices of hexagons and pentagons on spherical surfaces. A well-studied carbon-based nanomaterial is fullerene C60, which contains 60 carbon atoms located at the vertices of 12 pentagons and 20 hexagons. Its diameter is about 0.7 nm. Carbon nanotubes consist of rolled graphene sheets with a cylindrical shape and diameter of several nanometers. They are of different types depending on size, shape, and the total number of layers. Carbon nanotubes may be single- or double-walled nanotubes. Their thermal and electric properties are much better than those of other conductive materials.

Graphene is a bidirectional allotropic form that has carbon atoms with sp$_2$ hybridization linked to each other through the sigma and pai bond. Graphene possesses unique physical properties such as thermal stability and high rigidity; their electrical properties are different from materials with three dimensions [14, 15].

1.4.2 Metal-Based Nanomaterials

Metal-based nanomaterial consists of different types of metallic NP metal oxides such as titanium dioxide and quantum dots. Quantum dots are a closely packed group of crystals containing hundreds or thousands of atoms. Their optical properties depend upon their size. Metal-based nanomaterials have been extensively explored in the field of biomedical and pharmaceutics. The power of metal-based NPs lies in their ability to form a chemical bond with antibodies and pharmaceutics. They can be synthesized in a variety of ways, such as micro-emulsion, and by manipulating metal salt through chemical co-precipitation [16]. Metallic NPs consist of a central core of metal surrounded by shells of organic and inorganic compounds. Metallic NPs are toxic for living systems and their synthesis in powdered form is considered to be dangerous; this issue is resolved through the encapsulation of metallic NPs in hot film. Great research has been done on the synthesis of metallic NPs but there are still a lot of problems related to the impact of toxicity on human health and the environment. Because of their unique physicochemical properties, they have a vast range of applications in biomedicine and significant therapeutic effects. By investigating their way of interacting with plant and animal cells, their activities in health sciences to diagnose and treat several chronic disease demonstrate their therapeutic efficacy [17].

1.4.3 Semiconductor Nanomaterials

There is an increasing interest in the synthesis, characterization, and application of semiconductor nanomaterials in different fields. They have applications between conductors and non-conductors. A drastic change in physical and chemical properties occurs with a change in size, resulting in unique properties due to a large surface area. A semiconductor crystalline structure, a few nanometers in size and having specific physical and electrical properties, is termed a semiconductor nano-crystal [18]. Semiconductor nanomaterials are promising technological materials as their electronic and optical properties can be controlled and used to exploit their application, such as in bio-labels, light-emitting diodes, single-molecule transistors, and solar cells. Semiconductor nanomaterials or quantum dots are useful for the labeling of DNA, proteins, and cells. Their optical properties depend upon various parameters such as size, shape, and surface. The most common semiconductor material includes cadmium telluride, cadmium selenide, zinc sulfide, and zinc oxide. Semiconductor nano-crystals are made from different compounds; based on their position in groups of the periodic table they have been categorized into II-VI, III-V, and IV-VI semiconductor nano-crystals [19, 20].

1.4.4 NANOCOMPOSITE

The nanocomposite is a multiphase material, having nano-size particles embedded in a matrix of another material, resulting in the improvement of properties of that material such as strength, hardness, and electrical and thermal conductivity. The amount of nano-size material added to standard material is usually 0.5–5% by weight [21]. NPs cause drastic changes in the properties of bulk material as they have a large surface area-to-volume ratio; also they change how NPs attach with bulk material, resulting in improvement in properties. They have a broad range of applications in different fields; they are used in impellers and blades, thin-film capacitors for computer chips, food packaging, and oxygen and gas barriers. Nanocomposites also occur in nature, such as in the structure of bone. They have created great advances in the field of sensing technology. Polymer-based nanocomposites have been extensively used in this regard. The sensing properties of the sensor depend upon the type of interaction between the nanocomposite and analytic such as gas, chemical, or biological species [22]. Nanocomposite-based sensors have improved performance compared to other traditional sensors. Nanocomposite hydrogel has significant involvement in biomedical fields such as stem cell modification, tissue engineering, drug delivery, and medical devices. The great efficiency of hydrogel is due to the interaction between the polymer chain and nanocomposite. The main challenge in the synthesis of nanocomposites is the uniform dispersion of NPs because dispersion quality affects the interaction between phases and alters the final properties of nanocomposites. Nanocomposite has also been applied in osteointegration and osteoinductivity. There are two main aims in these fields: the first is to construct the nanocomposite with constituted material that will have good osteointegration with cartilage tissue and the second is to fabricate biomimetic nanostructures [23].

1.5 PHYSICAL SYNTHESIS OF NANOMATERIALS

1.5.1 THERMAL DECOMPOSITION OR THERMOLYSIS

There are various approaches for the synthesis of NPs, but thermal decomposition is the most innovative, easy, and economical one compared to other conventional methods. Thermal decomposition involves the use of heat for the chemical decomposition of the compound; it is an endothermal process. Heat is used to break the chemical bond in a compound and NPs are synthesized. This method was first developed by Hyeon, and despite various challenges it is still in practice [24]. The modern procedure is a modified version of the Hyeon method for the synthesis of various types of NPs. This method is cost-effective as it does not require expensive raw materials and complicated instruments. It involves the use of a precursor as a synthesis of copper NPs is done using copper salicylaldimine as a precursor. The precursor was prepared in powdered form and the powder morphology was determined using TEM. The powder was dissolved in ethanol. Then a solution was made containing a specific amount of copper salicylidimine and $C_{18}H_{37}N$ and solely heated at 130°C. Under nitrogen there was a slow change in color from red to green. The presence of oxygen and water created difficulties; the solution was kept at this temperature for 45 minutes and heated quickly. The solution was allowed to cool at room temperature and precipitated in ethanol. When exposed to air, newly synthesized NPs the color of the solution turned blue, which confirms the formation of NPs [25].

The temperature and pressure at which thermal decomposition occurs are greatly influenced by a reactive force with ligand in a coordination compound. Thermal decomposition does not involve the use of any stabilizer; its mean coordination compound can be thermostable. It has been observed that capping agents such as carboxylic acid and alkyl amine also influence the synthesis of NPs through thermal decomposition. Synthesis of NPs by thermal decomposition is becoming increasingly important because it is easy to handle the reaction conditions, particle size, structure, and purity. In addition to copper, cobalt NPs have been synthesized through the thermal decomposition method using bis(salicylidene)cobalt(II) as a precursor molecule. The resultant NPs have a diameter of 25–30 nm [25, 26].

1.5.2 LASER ABLATION METHOD

Laser ablation is the most versatile and promising method for the synthesis of various types of nanomaterials such as semiconductor nanodots, nanotubes, and NPs. It does not involve high temperature or any chemical reagent so it is considered to be environmentally friendly. In this method, a laser beam is allowed to interact with the target surface that may lie in a solid or liquid medium and NPs are collected in the form of a powdered or colloidal solution [27]. High-energy laser beams cause absorption of a large amount of light on the surface of the target, and as a result the temperature of the absorption material increases and the surface material is vaporized into a laser plume. In some cases, vaporized material is condensed into the cluster and, in some cases, it reacts with other introduced substances to produce new material. Newly synthesized nanomaterials are then collected from the substrate surface. The NPs produced do not contain any insulating capping agent that made their use possible in organic electronics [28]. This method allows the use of the ligand of choice and synthesis of NPs in the medium of choice. Another significant advantage of this method is that nanomaterials with desired properties such as size, shape, structure, and composition can be produced as these depend upon the laser beam parameter, such as pulse energy, rate, and repetition rate. Moreover, the synthesis of NPs in a liquid medium offers the advantage of adjusting the size of NPs according to need simply by adjusting ablation time and irradiating the final colloidal solution. Synthesis of nanomaterials in the air is different from a liquid medium; liquid may impose physical and chemical effects such as oxidizing, reducing, and cooling effects which affect the rate of ablation. Production and distribution of nanomaterials are higher when laser ablation is performed on a metal plate in a liquid medium. The size of NPs that can be obtained in this method is 10 nm [28, 29].

1.5.3 RADIOFREQUENCY PLASMA METHOD

This is a novel and useful method, especially for biomedical material that is not compatible with certain types of chemicals. NPs have been synthesized through the radiofrequency (RF) plasma method. There are three basic components of this method: (1) an ultrasonic spray for generating aerosol; (2) an RF initiation system to induce the decomposition of aerosol; and (3) a collection system [30]. The initial substance or metal from which NPs are to be synthesized is taken in a pestle and placed in an evacuated chamber. RF coils are wrapped around the chamber; these coils heat the metal to above its evaporation point. There is a hole on one side of the chamber through which helium gas is allowed to enter the chamber. Helium gas produces plasma in the region of coils. Metal vapors interact and nucleate on the atoms of helium gas and move towards the upper region. There is a cold collector rod in the upper region where NPs are collected [31]. The RF thermal plasma method is very useful because of the high chemical reactivity due to high temperature, its capability of producing NPs of the same size, and its high production rate. In this method the size of particles depends upon the initial vapor concentration and cooling rate. This process is usually conducted at atmospheric pressure. Temperature gradient near plasma mainly contributes to NP synthesis. Another significant advantage of this method is the contamination-free synthesis of NPs as it avoids the use of internal electrodes. In the gas phase RF plasma has been used to synthesize metal oxide NPs. However, it is expected that RF plasma in the liquid phase can also be used to synthesize different types of NPs although little research work has yet been done; there is a need for more investigation. Synthesis of some NPs through this method has been reported in previous studies and it was found that gold NPs cannot be synthesized through this method [32].

1.6 CHEMICAL SYNTHESIS OF NANOMATERIALS

1.6.1 CO-PRECIPITATION METHOD

The co-precipitation method is a wet chemical method that is also known as the solvent displacement method. In this method, the polymer phase can be either natural or synthetic and the polymer solvents are hexane, ethanol, acetone, and non-solvent polymer. During this method, NPs are rapidly formed by diffusion of polymer solvent into a non-solvent polymer by mixing the solvent and non-solvent polymer solution at the end. NPs are formed by interfacial tension at two polymer phases [33].

1.6.2 SOL-GEL METHOD

This method involves the thermal decomposition, condensation, and hydrolysis of metal precursors or alkoxides and forms a stable solution called sol which upon condensation or hydrolysis forms a gel that increases viscosity. The size of the particle in the gel can be observed by changing precursors' pH values, temperature, and concentration. After that, the development of solid mass is mandatory; this might take a few days. During that period removal of solvent, phase alteration, and Ostwald ripening could take place. The unstable reagent is separated to produce NPs [34, 35].

1.6.3 MICROWAVE-ASSISTED SYNTHESIS

Microwave-assisted methods have been used on a large scale for the synthesis of oxides, sulfide, and hydroxide NPs. This method is simple, environmentally friendly, and without the problems of thermal gradient effects. It has a lot of advantages in comparison to convenient hydrothermal methods, such as a simple medium, quick reaction, less time to attain a suitable temperature for reaction, and the morphology of particles is also under control. Microwave-assisted synthesis of NPs has the significant advantage of fast and homogeneous heating of precursor materials. Radiation released in the microwave has penetration characteristics which help in heating the reaction solution evenly, leading to the formation of crystallites with small size distribution [36].

1.6.4 SONOCHEMICAL METHOD

Nano-hybrids such as palladium and copper oxide (Pd-CuO) have been efficiently made using the sonochemical method by the fusion of Cu salt in the presence of Pd and water. In this method, metal salts are converted into metal oxides in the presence of palladium and water with ultrasound energy. Moreover, the Pd is used in the form of palladium salts or pure metallic palladium [37].

1.7 BIOLOGICAL SYNTHESIS OF NANOMATERIALS

1.7.1 PLANT-BASED SYNTHESIS

Nanotechnology is a rapidly growing field with a wide range of applications in medicine and modern science. There are various physical and chemical methods for the synthesis of NPs, but they involve the use of chemicals and high energy demand. Recently plant-based synthesis of the NP has been gaining the attention of researchers and is widely used. It is cost-effective, rapid, and environmentally friendly. Plant-based synthesis of NPs began at the start of the 20th century. Plants contain different phytochemicals with high therapeutic potential [38]. Plant-based synthesis was first investigated for gold and silver NPs. Extracts of different parts of plants [oat (*Avena sativa*), neem (*Azadirachta indica*), alfalfa (*Medicago sativa*), tulsi (*Ocimum sanctum*), lemon (*Citrus limon*), and mustard (*Brassica jounce*) were investigated for the synthesis of gold and silver NPs. There are various parameters that are effective in the green synthesis of NPs, such as the use of suitable solvent, temperature, pressure, and pH. Plant-based synthesis is considered to be a successful method because of the potential of plants to accumulate different heavy metals in their parts [39]. Various plants have been investigated for their potential to reduce metallic salt for the synthesis of NPs. It is reported that *Aloe vera* leaf extract has been used for the synthesis of selenium NPs. The presence of various secondary metabolites such as phenolic compounds, organic acid, polysaccharides, and vitamins was detected in the leaf extract. It was found that these metabolites act as natural reducing and stabilizing agents for the reduction of selenium salt. However, the synthesis of selenium NPs has also been reported from various other plants, such as *Vitis vinifera*, *Allium sativum*, *Carica papaya*, and fruit extract of *Zingiber officinale* [40].

1.7.2   FUNGAL-BASED SYNTHESIS

Fungal-based synthesis of different metallic NPs is considered to be an efficient method as fungi possess high tolerance against different metals. They secrete various types of extracellular proteins that are involved in the stabilization of NPs. Compared to plant-based synthesis, fungi offer more advantages as they are resistant to agitation and pressure. NPs with desired characters can be obtained by controlling the culture conditions such as pH, temperature, time, and amount of biomass [41]. It is easy to grow and maintain a large culture of fungi; they produce a variety of compounds with applications in different fields. Synthesis of NPs from fungi may comprise extracellular or intracellular mechanisms. Intracellular mechanisms involve the addition of metal precursors within the fungal culture. Extraction of NPs is done after synthesis using different techniques such as centrifugation, different chemical treatments, and filtration. In the extracellular method, the precursor is added to a dispersion containing fungal biomolecules, and NPs are formed. Synthesis of CuNPs from fungal biomass is reported, and fungal isolates were obtained from soil, rotten fruits, etc. [42]. Petri plates were washed and sterilized for culturing fungus. Subcultures were obtained to get purified fungal isolates. Then biomass was prepared by inoculating fungal spores in Sabouraud dextrose broth and incubated at 35°C. It was placed in an arbitrary shaker for 12–13 days, and then biomass was filtered and washed with distilled water. Then the obtained filtrate was mixed with the different concentrations of $CuSO_4$ (metal salt) in a ratio of 1:1. The final yield of NPs is measured using a spectrophotometer. Besides this synthesis of silver and gold NPs has been done using fungi such as *Penicillium*, *Fusarium*, *Trichoderma*, and *Aspergillus*. The optimal conditions for the synthesis of AgNPs from fungi are 7 pH, 25°C temperature, 1 mM silver nitrate, and 15–20 g wet cell filtrate. NPs obtained under these conditions are usually spherical and well dispersed [43].

1.7.3   BACTERIAL-BASED SYNTHESIS

Bacteria are considered to be potential candidates for the synthesis of different types of NP as they possess an outstanding ability to reduce metals. Bacterial-based synthesis of nanomaterials is the prominent emerging field. Some bacterial species are reported to have a mechanism for tolerating heavy-metal stress and surviving at high metal concentrations. Some bacterial species, such as *Thiobacillus ferrooxidans*, *T. thiooxidans*, and *Sulfolobus acidocaldarius*, can reduce ferric ions to ferrous as they use elemental sulfur as an energy source. Synthesis of AgNPs has been done using the culture of *Bacillus subtilis* [44]. To increase the rate of synthesis and inhibit the aggregation of NPs microwave radiation was used. This radiation provides uniform heating

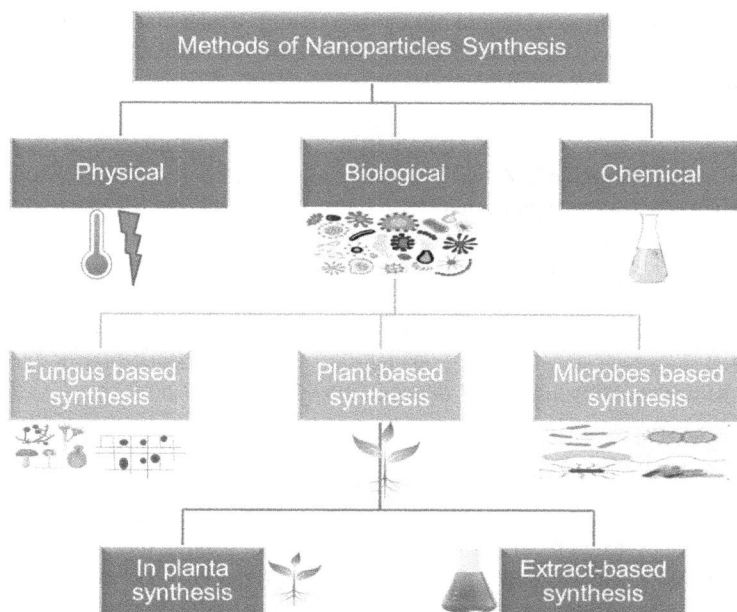

FIGURE 1.1 Routes for the synthesis of nanomaterial.

around NPs. Synthesis of silver NPs has been done using a culture supernatant of *Bacillus licheniformis*. Well-dispersed silver nano-crystals were obtained and were stable in an aqueous solution at room temperature when stored in the dark. These NPs also exhibit antibacterial properties against various drug-resistant microorganisms. It has been found that the stability of synthesized silver NPs depends upon the temperature, pH, and type of bacterial culture used. *Phaeocystis antarctica* synthesized silver NPs at a particular temperature but no synthesis was observed for *Arthrobacter kerguelensis* at the same temperature. Besides this, the synthesis of gold NPs has been carried out using a culture of *Escherichia coli*. Synthesized NPs were not homogeneous in size and shape; some were spherical and triangular. Through microscopic studies, they were found on the surface of bacteria. In another study, gold was recovered from waste of jewelry shops using a culture of *E. coli.* [45]. An illustration of various routes for the synthesis of nanomaterial is presented in Figure 1.1.

1.8 CHARACTERIZATION OF NANOMATERIALS

1.8.1 UV Spectrophotometer

UV-visible spectroscopy is the most reliable and widely used technique for the structural characterization of NPs. It measures the amount of light absorbed and scattered by the sample. A sample is positioned between a light source and photodetector and the intensity of a beam of light is measured by a spectrophotometer after it passes through a sample. Light absorption is measured at different wavelengths, and then these measurements are compared to find out the wavelength-dependent extinction spectrum of the sample. To ensure that spectral characteristics from the solvent are not included in the sample extinction spectrum, each spectrum is background-corrected using a "blank," which is a cuvette filled with the dispersing media [46]. This technique is also used to diagnose the formation of NPs and monitor their stability. Unique optical properties of silver and gold NPs make UV-visible spectroscopy an important tool for identifying, characterizing, and investigating nanomaterials and causing them to interact strongly with specific light wavelengths. The absorption spectra of NPs show a highly specific symmetric band with characteristic peaks; for example, the absorption band of silver NPs shows a characteristic peak at 390–470 nm. The peak within this range confirms the synthesis of NPs. Characteristic peaks of absorption spectra vary as NPs transition from a well-dispersed state to an aggregated state. UV-visible spectroscopy can be used to find whether the NP solution has destabilized over time [47].

1.8.2   FOURIER TRANSFORM INFRARED (FT-IR) SPECTROSCOPY

FT-IR spectroscopy is a widely used analytical technique for determining the molecular composition and structure of an unknown sample. In this technique, infrared radiation is allowed to pass through the sample, and the range of wavelength absorbed by the sample in the infrared region is measured. In the FT-IR instrument, infrared radiation in the range of 10,000–100 cm^{-1} is allowed to pass through the sample; some is absorbed by the sample, and some radiation passes through it [48]. Radiation absorbed by the sample is converted into vibrational energy of the sample molecule and finally reaches the detector and appears as a spectrum. It represents each molecule and chemical structure present in the sample. It is a useful method in the surface characterization of NPs. For FT-IR measurements of NPs, the colloidal solution of particles is allowed to dry in a desiccator, and a spectrum was attained within the range of 4000–400 cm^{-1}. Under specific conditions, not only functional groups attached to the surface of NPs are determined but different active sites responsible for the surface reactivity of NPs are also determined. The coordination of different capping agents on the surface of NPs, particularly silver and gold NPs, has been determined [49]. FT-IR spectrum shows a characteristic absorption band corresponding to the presence of the particular functional group. For example, in one study, the FT-IR spectrum of silver NPs and that of precursor ($AgNO_3$) were compared, and an intense absorption band was found to be present in the spectrum of $AgNO_3$, which is a characteristic of $Ag^+NO_3^-$. In contrast, in the spectrum of silver NPs this band was shifted to a lower-wavenumber value due to interaction with the silver NP. Similarly, peaks at different wavelengths further confirm the formation of NPs. Besides this, FT-IR spectroscopy is used in controlling the quality of industrial products. A change in a particular pattern of absorption band indicates a change in the composition of the material and the presence of contaminants. It is a useful technique for determining the chemical properties of the material as small as 10–15 microns [50].

1.8.3   ATOMIC FORCE MICROSCOPY (AFM)

AFM is the most useful microscopic technique to study and analyze samples at the nanoscale. It not only creates three-dimensional images but also provides important information about the surface characteristic of the sample and produces an image at high resolution with an angstrom scale. AFM consists of the following parts that help to perform its function: a modified tip is used to detect the presence of the sample, and software to adjust the image. It works on a laser beam deflection system. A laser beam is deflected from the back of the AFM lever towards the detector and an image is produced. It is highly dependent on force measurement because of its important contribution to image production. It operates in two modes, tip contact and tapping mode. In contact, the mode tip comes in contact with the sample while in tapping mode its cantilever moves above the sample surface, and the tip comes in contact with the sample occasionally [51].

AFM has various advantages as it is easy to prepare a sample to observe in AFM. It measures sample size accurately and produces a three-dimensional image. It can analyze both living and non-living material and measure the roughness of the surface. The outstanding role of AFM in the characterization of NPs makes it an excellent tool for determining size distribution and producing a high-quality image. NPs act as calibration standards in AFM. In this instrument, vertical resolution is less than 0.1 nm and x–y resolution is about 1 nm.

Another outstanding ability of AFM is that it can characterize the sample in a variety of media, such as air, controlled environment, and liquid dispersion. It creates informative topographies of NPs and sometimes, in the case of composite samples, it can distinguish between different particles by producing spatial distribution information. In comparison to scanning electron microscopy (SEM) and TEM, it is a cost-effective microscopic technique for nanoscale measurements. It requires less space and there is no need for specially trained operators. AFM not only creates a three-dimensional image but also provides quantitative information about the height of NPs [52].

1.8.4 Transmission Electron Microscopy

TEM is a high-magnification measurement technique used in the characterization of NPs. It allows the transmission of an electron beam through a sample and an image is formed after its pass through the sample. The image is magnified and focused on the objective lens, and ultimately it appears on the screen. TEM has extremely high resolution because it uses a beam of an electron rather than a beam of light to illuminate the sample. TEM is a highly valuable technique to find out the particle size, size distribution, and morphology of NPs [46]. It is very important to find out the size and size distribution of NPs before applying them in any field, so, TEM can be the best way to find the size and shape of NPs, and an image formed on screen confirms the formation of NPs. It is also becoming the preferred technique to explore the effect of NPs on a biological system. To play their role in therapeutics, NPs must interact with the biological system. Due to its high resolution, TEM provides valuable information on the whole uptake mechanism. An uptake mechanism plays an important role in determining the intracellular fate of NPs. It has also been observed that NPs adhere to each other through electrostatic and other interactions which affect their ability to perform their function inside the cell. TEM can be the best tool for getting information on these interactions and providing a suitable method for sample fixation [53, 54].

1.8.5 SCANNING ELECTRON MICROSCOPY

SEM is a versatile technique for analyzing microstructure morphology and chemical composition characterization. It uses a beam of electrons for scanning the sample. A beam of electrons is fired through an electron gun and then accelerated down towards the column of a scanning electron microscope [46]. During this progression, it passes through a series of lenses. The whole process occurs in vacuum conditions to prevent the interaction of atoms already present in the column of the microscope with a beam of electrons and to produce a high-quality image. SEM is a promising device for the characterization of NPs. It directly observes NP dimensions and measures geometric size directly in the IS length unit, the meter. Due to this reason, SEM is considered a powerful tool in the characterization of the size, size distribution, and shape of NPs. Sample preparation is a key step in this process; if there is an agglomeration state it will negatively affect the measurement results of constituent NPs. Dispersion of NPs on substrate reduces the chances of error in measurements. It does not require any mechanical crushing of the sample, thus morphological and surface characteristics of the sample are retained [53, 54].

1.8.6 VIBRATING SAMPLE MAGNETOMETER (VSM)

A VSM is a scientific instrument used to study the magnetic behavior of a sample. It works on Faraday's law of induction, which explains that changing a magnetic field creates an electric field and this electric field provides information about a changing magnetic field. To measure the magnetic properties, the sample is placed in VSM and a constant magnetic field is applied. If the sample possesses magnetic properties, then this field will magnetize the sample by creating a magnetic dipole with the field. Due to this magnetic dipole the sample will move and this movement will create a magnetic field around the sample, called a stray magnetic field [55]. This magnetic field changes with time when samples move up and down and can be measured by pick-up coils. This changing magnetic field induces an electric field in pick-up coils according to Faraday's law of induction, as described previously. This electric current indicates the magnetization of the sample: the greater the magnetization, the greater will be the current in the coils. VSM has also been used to study the magnetic behavior of NPs. It indicates the magnetization of NPs by showing a magnetization loop. Material that gives a small hysteresis curve is considered to have a low value of coercivity and retentivity. Results of VSM show that the value of coercivity (the field needed to magnetize the material) is 61.127. And the value of retentivity (a field required to demagnetize the material) is 5.4324 emu/g. According to data obtained from measurements of VSM, the value of magnetization for NPs is 70.787 emu/g [56].

1.8.7 ENERGY-DISPERSIVE X-RAY SPECTROSCOPY (EDS)

EDS is the best-known method to analyze the elemental composition of nano-sized material. The EDS analysis of any sample requires only an X-ray associated with the equipment of SEM. It focuses the beam of charged particles, such as an electron and protons, on to the sample. This beam strikes the electron in the inner shell and expels it out, creating a hole; the electron from the outer shell fills this hole. The difference in energy between two shells causes the emission of X-rays. X-rays emitted from the sample are measured through the EDS [57]. As the energy of two X-rays is related to the energy difference between two shells of an atom in the sample, it allows the elemental composition of the sample to be measured. It has become the most powerful technique due to its use in determining the elemental composition of NPs in association with SEM. In various studies, different types of NPs such as gold and silver were synthesized and their formation was confirmed through EDS analysis, which showed the presence of intense bands in the spectrum. The detection limit of EDS varies according to the surface condition of the sample. If there is a smooth surface, then the detection limit will be lower. This technique plays a significant role in drug delivery as it is used to detect the presence of NPs. It is also used to study environmental pollution as it indicates the presence of heavy metals. Different peaks in the EDS spectrum give qualitative as well as quantitative information as the position of the peak indicates the type of element present in the sample and the area of the peak designates the amount of element in the sample [57, 58].

1.8.8   X-ray Photoelectron Spectroscopy (XPS)

XPS is a simple quantitative technique used to determine the elemental composition of nanoscale material; it also indicates the binding state of elements. In comparison to many other techniques, XPS is unique in its utility, versatility, and popularity. It not only detects the presence of certain elements but also finds the species of the same element as a change in chemical bond alters the electronic configuration of an atom [59]. The sample must first be evacuated before processing in XPS. XPS is based on the simple process that a photon having particular energy is passed through the surface of the sample, and when the kinetic energy of the electron increases, it ejects out from the surface. The kinetic energy is analyzed after being emitted from the surface. It gives information on the electronic state of atoms present in the surface region of the sample. In XPS a spectrum is plotted between a relative number of electrons and their binding energies. Shorter peaks show fewer electrons. For example, if peak 1 is half the height of peak 2, it means the number of electrons detected at peak 1 is half the number of electrons detected with binding energy at peak 2. It has been proved to be the most useful technique in the elemental analysis of NPs. It offers more advantages in comparison to other techniques. It provides highly reliable quantitative information on the chemical composition of nanomaterials. It has been used to determine the functional groups on the surface of the nano-complex made of self-assembled NPs for controlling drug delivery. It also gives information on the possible interactions of nano-complexes with their surroundings [60].

1.8.9   Magnetic Force Microscopy (MFM)

MFM is a scanning technique used to study the surface and magnetic properties of the sample at a nanoscale. This technique involves the use of a magnetic probe that moves over the sample surface and interacts with its magnetic field. It is commonly called the two-pass method as it scans the sample surface twice [61]. The first time it produces a topographical image and the second time it produces an MFM image. It is considered a powerful tool since it can evaluate the magnetic properties of even a single NP. It is mostly used for high-resolution images and to study molecular interactions. It is widely used in the study of a sample at the nanoscale due to the combination of two unique properties: one is high-resolution image-forming ability and the second is high sensitivity to the nearby magnetic field. This technique has been used to spot the biomolecules in the biological system by conjugating them with magnetic NPs. It is a promising technique and is considered an alternative to SEM for studying the uptake of different types of NPs by cell. Another advantage is that it requires minimal sample preparation. It has been used to study human cancer cell lines by labeling them with iron oxide NPs [62].

1.9 CONCLUSIONS

Nanomaterials are a class of materials that are manufactured by different techniques at the nanoscale level (1–100 nm). Recently, nanomaterials have become one of the most powerful tools in science and technology globally. The unique features of nanomaterials enable them to have potential applications in science and industry. This chapter provides an overview of the history of nanomaterials and their different features and characterization techniques, as well as their various fabrication methods. It also illustrates novel research and development in nanomaterials.

REFERENCES

1. Darweesh HHM. Nanomaterials: classification and properties – part I. Journal of Nanoscience, 2018; 1(1): 1–11.
2. Włodarczyk R, Kwarciak-Kozłowska A. Nanoparticles from the cosmetics and medical industries in legal and environmental aspects. Sustainability, 2021; 13(11): 5805.
3. Mandal A, Ray Banerjee E. Introduction to nanoscience, nanotechnology and nanoparticles. *In*: Nanomaterials and Biomedicine. 2020; Springer, Singapore. pp. 1–39.
4. Barhoum A et al. Review on natural, incidental, bioinspired, and engineered nanomaterials: history, definitions, classifications, synthesis, properties, market, toxicities, risks, and regulations. Nanomaterials, 2022; 12(2): 177.
5. Vahabi S, Mardanifar F. Applications of nanotechnology in dentistry: a review. General Dentistry, 2014; 32(4):228–239.
6. Dong J, Zhang J. Maya blue pigments derived from clay minerals. *In*: Nanomaterials from Clay Minerals. 2019; Elsevier, Amsterdam. pp. 627–661.
7. Patra JK, Gouda S. Application of nanotechnology in textile engineering: an overview. Journal of Engineering and Technology Research, 2013; 5(5): 104–111.
8. Sudha PN et al. Nanomaterials history, classification, unique properties, production and market. *In*: Emerging Applications of Nanoparticles and Architecture Nanostructures. 2018; Elsevier, Amsterdam. pp. 341–384.
9. Ganguly P et al. 2D nanomaterials for photocatalytic hydrogen production. ACS Energy Letters, 2019; 4(7): 1687–1709.
10. Khan I, Saeed K, Khan I. Nanoparticles: properties, applications and toxicities. Arabian Journal of Chemistry, 2019; 12(7): 908–931.

11. Nam S et al. Enhancement of electrical and thermomechanical properties of silver nanowire composites by the introduction of nonconductive nanoparticles: experiment and simulation. ACS Nano, 2013; 7(1): 851–856.

12. Manna I. Synthesis, characterization and application of nanofluid — an overview. Journal of the Indian Institute of Science, 2009; 89(1): 21–33.

13. Rajak DK et al. Recent progress of reinforcement materials: a comprehensive overview of composite materials. Journal of Materials Research and Technology, 2019; 8(6): 6354–6374.

14. Georgakilas V et al. Broad family of carbon nanoallotropes: classification, chemistry, and applications of fullerenes, carbon dots, nanotubes, graphene, nanodiamonds, and combined superstructures. Chemical Reviews, 2015; 115(11): 4744–4822.

15. Pradhan B, Srivastava SK. Synergistic effect of three-dimensional multi-walled carbon nanotube–graphene nanofiller in enhancing the mechanical and thermal properties of high-performance silicone rubber. Polymer International, 2014; 63(7):1219–1228.

16. Hedayatnasab Z, Abnisa F, Daud WMAW. Review on magnetic nanoparticles for magnetic nanofluid hyperthermia application. Materials & Design, 2017; 123: 174–196.

17. Saratale RG et al. New insights on the green synthesis of metallic nanoparticles using plant and waste biomaterials: current knowledge, their agricultural and environmental applications. Environmental Science and Pollution Research, 2018; 25(11): 10164–10183.

18. Chen J et al. Wet-chemical synthesis and applications of semiconductor nanomaterial-based epitaxial heterostructures. Nano-Micro Letters, 2019; 11(1): 1–28.

19. Biju V et al. Semiconductor quantum dots and metal nanoparticles: syntheses, optical properties, and biological applications. Analytical and Bioanalytical Chemistry, 2008; 391(7): 2469–2495.

20. Houtepen AJ et al. On the origin of surface traps in colloidal II–VI semiconductor nanocrystals. Chemistry of Materials, 2017; 29(2): 752–761.

21. Camargo PHC, Satyanarayana KG, Wypych F. Nanocomposites: synthesis, structure, properties and new application opportunities. Materials Research, 2009; 12(1): 1–39.

22. Bharadwaz A, Jayasuriya AC. Recent trends in the application of widely used natural and synthetic polymer nanocomposites in bone tissue regeneration. Materials Science and Engineering C, 2020; 110: 110698.

23. Kumar SK, Krishnamoorti R. Nanocomposites: structure, phase behavior, and properties. Annual Review of Chemical and Biomolecular Engineering, 2010; 1: 37–58.

24. Darezereshki E et al. A novel thermal decomposition method for the synthesis of ZnO nanoparticles from low concentration $ZnSO_4$ solutions. Applied Clay Science, 2011; 54(1): 107–111.

25. Betancourt-Galindo R et al. Synthesis of copper nanoparticles by thermal decomposition and their antimicrobial properties. Journal of Nanomaterials, 2014; 1–5. https://doi.org/10.1155/2014/980545

26. Hufschmid R et al. Synthesis of phase-pure and monodisperse iron oxide nanoparticles by thermal decomposition. Nanoscale, 2015; 7(25): 11142–11154.

27. Naser H et al. The role of laser ablation technique parameters in synthesis of nanoparticles from different target types. Journal of Nanoparticle Research, 2019; 21(11): 1–28.

28. Zhang D, Gokce B, Barcikowski S. Laser synthesis and processing of colloids: fundamentals and applications. Chemical Reviews, 2017; 117(5): 3990–4103.

29. Mahdieh MH, Fattahi B. Size properties of colloidal nanoparticles produced by nanosecond pulsed laser ablation and studying the effects of liquid medium and laser fluence. Applied Surface Science, 2015; 329: 47–57.

30. Jankeviciute A et al. Synthesis and characterization of spherical amorphous alumo-silicate nanoparticles using RF thermal plasma method. Journal of Non-Crystalline Solids, 2013; 359: 9–14.
31. Avramescu SM et al. Engineered Nanomaterials: Health and Safety. 2020; IntechOpen, London. 10.5772/intechopen.83105.
32. Zhang H et al. Single-step pathway for the synthesis of tungsten nanosized powders by RF induction thermal plasma. International Journal of Refractory Metals and Hard Materials, 2012; 31: 33–38.
33. Das S, Srivasatava VC. Synthesis and characterization of ZnO–MgO nanocomposite by co-precipitation method. Smart Science, 2016; 4(4): 190–195.
34. Mackenzie JD, Bescher EP. Chemical routes in the synthesis of nanomaterials using the sol–gel process. Accounts of Chemical Research, 2007; 40(9): 810–818.
35. Li J, Wu Q, Wu J. Synthesis of nanoparticles via solvothermal and hydrothermal methods. In: Handbook of Nanoparticles. 2016; Springer, Cham. DOI: 10.1007/978-3-319-15338-4_17
36. Horikoshi S, Serpone N. Microwave-assisted synthesis of nanoparticles. In: Microwave Chemistry. 2017; De Gruyter, Berlin. pp. 248–269.
37. Ziylan-Yavas A et al. Supporting of pristine TiO$_2$ with noble metals to enhance the oxidation and mineralization of paracetamol by sonolysis and sonophotolysis. Applied Catalysis B: Environmental, 2015; 172: 7–17.
38. Ikram M et al. Biomedical potential of plant-based selenium nanoparticles: a comprehensive review on therapeutic and mechanistic aspects. International Journal of Nanomedicine, 2021; 16: 249.
39. Alabdallah NM, Hasan MM. Plant-based green synthesis of silver nanoparticles and its effective role in abiotic stress tolerance in crop plants. Saudi Journal of Biological Sciences, 2021; 28(10): 5631–5639.
40. Ikram M et al. Foliar applications of bio-fabricated selenium nanoparticles to improve the growth of wheat plants under drought stress. Green Processing and Synthesis, 2020; 9(1): 706–714.
41. Priyadarshini E et al. Metal–fungus interaction: review on cellular processes underlying heavy metal detoxification and synthesis of metal nanoparticles. Chemosphere, 2021; 274: 129976.
42. Jampílek J, Králová K. Impact of nanoparticles on toxigenic fungi. In: Nanomycotoxicology. 2020; Elsevier, London. pp. 309–348.
43. Banerjee K, Ravishankar Rai V. A review on mycosynthesis, mechanism, and characterization of silver and gold nanoparticles. Bio Nano Science, 2018; 8(1): 17–31.
44. Yu X et al. Green synthesis and characterizations of silver nanoparticles with enhanced antibacterial properties by secondary metabolites of Bacillus subtilis (SDUM301120). Green Chemistry Letters and Reviews, 2021; 14(2): 190–203.
45. Aftab A. Synthesis of nanoparticles by microbes. In: Nanobotany. 2018; Springer, Cham. pp. 175–193.
46. Kumar PS, Pavithra KG, Naushad M. Characterization techniques for nanomaterials. In: Nanomaterials for Solar Cell Applications. 2019; Elsevier, Amsterdam. pp. 97–124.
47. Metin CO et al. Stability of aqueous silica nanoparticle dispersions. Journal of Nanoparticle Research, 2011; 13(2): 839–850.
48. Smith BC. Fundamentals of Fourier Transform Infrared Spectroscopy. 2011; CRC Press, Boca Raton, FL.
49. Aguilar-Méndez MA et al. Synthesis and characterization of silver nanoparticles: effect on phytopathogen Colletotrichum gloesporioides. Journal of Nanoparticle Research, 2011; 13(6): 2525–2532.

50. Ahmad T et al. Biosynthesis, structural characterization and antimicrobial activity of gold and silver nanoparticles. Colloids and Surfaces B: Biointerfaces, 2013; 107: 227–234.

51. De Oliveira R et al. Measurement of the nanoscale roughness by atomic force microscopy: basic principles and applications. *In*: Atomic Force Microscopy: Imaging, Measuring and Manipulating Surfaces at the Atomic Scale. 2012; InTechOpen, London.

52. Mishra R et al. The production, characterization and applications of nanoparticles in the textile industry. Textile Progress, 2014; 46(2): 133–226.

53. Mielańczyk Ł et al. Transmission electron microscopy of biological samples. *In* The Transmission Electron Microscope: Theory and Applications. 2015; IntechOpen, London. pp. 193–239.

54. Inkson B. Scanning electron microscopy (SEM) and transmission electron microscopy (TEM) for materials characterization. *In*: Materials Characterization Using Nondestructive Evaluation (NDE) Methods. 2016; Elsevier, Duxford. pp. 17–43.

55. Ayyappan S, Philip J, Raj B. A facile method to control the size and magnetic properties of $CoFe_2O_4$ nanoparticles. Materials Chemistry and Physics, 2009; 115(2–3): 712–717.

56. Lopez-Dominguez V et al. A simple vibrating sample magnetometer for macroscopic samples. Review of Scientific Instruments, 2018; 89(3): 034707.

57. Pytlakowska K et al. Determination of heavy metal ions by energy dispersive X-ray fluorescence spectrometry using reduced graphene oxide decorated with molybdenum disulfide as solid adsorbent. Spectrochimica Acta Part B: Atomic Spectroscopy, 2020; 167: 105846.

58. Shindo D, Oikawa T. Analytical Electron Microscopy for Materials Science. 2002; Springer Science & Business Media, Berlin.

59. Hollander JM, Jolly WL. X-ray photoelectron spectroscopy. Accounts of Chemical Research, 1970; 3(6): 193–200.

60. Wu CK et al. Quantitative analysis of copper oxide nanoparticle composition and structure by X-ray photoelectron spectroscopy. Chemistry of Materials, 2006; 18(25): 6054–6058.

61. Kazakova O et al. Frontiers of magnetic force microscopy. Journal of Applied Physics, 2019; 125(6): 060901.

62. Dadfar SM et al. Iron oxide nanoparticles: diagnostic, therapeutic and theranostic applications. Advanced Drug Delivery Reviews, 2019; 138: 302–325.

2 Nanomaterial as Nano-Pesticides

Devendra Kumar Verma, Rashmi Rameshwari, and Santosh Kumar

CONTENTS

2.1 INTRODUCTION

Nanomaterials have great power to disrupt the strength of other substances [1–6]. At a nanoscale level, quantum effects can become much more important in defining the material properties and characteristics that generate novel optical, electrical and magnetic behaviours. Currently, nanomaterials are being formed on an industrial scale,, although others are still involved in research and development and are formed on a smaller scale. Polymers and liposomes are mostly used as organic carriers to encapsulate pesticides as they are biodegradable; hence they are environmentally friendly, safe and renewable [7]. To pesticides, nanomaterials provide a larger specific surface area. Furthermore, as unique agrochemical carriers, nanomaterials enable crop protection by delivering site-targeted delivery of nutrients. Nano-pesticides come in a variety of formulations, such as fibre capsules, gels, etc. Nanoparticles can also serve as nano-pesticides on their own [8].

Currently nanomaterials are applied in huge amounts, particularly in agriculture, to get rid of weeds, fungi and insects. Due to this practice the accumulation of

DOI: 10.1201/9781003322122-2

nanomaterials in the environment is a major concern. More research is thus required to find out the effect of nano-pesticides on the environment. Nano-pesticides provide many benefits to the agriculture sector. Presently the risks and benefits of nano-pesticide are poorly known. Many researchers are actively involved to find their applications and in future will provide all the answers related to risk and benefits [9].

However, data related to the concentration of nanomaterials are lacking. Thus, concentrations in the order of nanograms are even lower. Studies have shown that nanomaterials based on carbon show toxicity at high concentrations [10]. Nanomaterials, such as nano-pesticides, may also have a positive impact on biodiversity. At extremely low concentrations they can have a positive effect on biodiversity [11]. Researchers have discovered that there are many pesticides which causes heresies with both positive as well as negative effects [12]. There is a need to consider nanomaterials and pesticides in ecological risk assessments. This has received little attention from researchers. At low concentrations, nanomaterials and pesticides individually induce hormesis [13]. Various studies have concluded that hormesis should be incorporated by considering its mode of action [14]. It can present great solutions in agriculture industry when integrated with green chemistry and applied to field studies [15]. Toxicity can be reduced by changing the method of preparation. Environmental factors can combine with pollutants and increased toxicity [16]. It has been observed that nanomaterial-induced toxicity can change soil fertility. To access this, various generations of crop data need to be accessed. Due to a lack of scientific evidence in this area the effect of nanomaterial toxicity has not yet been explored [17]. There is a need for modifications of agricultural policies through which farmers can benefit [18].

2.2   WHAT IS A NANO-PESTICIDE?

At the moment, the definition of "nano-pesticides" is not fixed. It is used differently in legislation, science, the public and business. According to suggestions by eminent scientists, "a nanoparticle is a plant protection product that contains components in the nanometer size range (up to 1000 nm) and has new properties that are related to the small size of their components" [19]. In contrast to other nanomaterials, definition based on size is insufficient for nano-pesticides [19]. Pesticides are always made up of several ingredients. An active ingredient (AI) is combined with other compounds, much like a drug. These enable the potent ingredient to be easily applied to the plant, evenly distributed and stable after use. The nanomaterial itself is not the active agent in most proposed applications for nano-pesticides but rather acts as a subsidiary compound to make the potent agent constant and enhance its controlled release (e.g. nanocapsule). Nanomaterials as active agents include nanocopper and nanosilver [20]. Nanomaterial use is based on certain benefits. In pesticides they are used in smaller amounts to increase efficacy. Research is still ongoing and these products are not currently widely applied in the field [21].

2.3   RISKS OF NANO-PESTICIDES FOR THE ENVIRONMENT

Because of the nature of their application, nanomaterials in nano-pesticides are purposely released into the environment. The benefits and risks for human health must be assessed [22]. According to EU regulations on biocidal products, it is compulsory for manufacturers to ensure the safety of nano-pesticides before farmers are allowed to use them. Presently many ideas related to nano-pesticide use have been floated. There is not yet such a nano-pesticide product available on the market. Developing and approving a pesticide is a time-consuming and expensive process; it is currently unknown how much money can be saved by using nano-pesticides instead of conventional pesticides [23].

2.4   BENEFITS OF NANO-PESTICIDES

Many ingredients that are considered AIs in pesticides are not properly soluble in water, making their application difficult. As accompanying substances, the ingredients are soluble in the presence of nanomaterials. This allows for better control of release and distribution, as well as better protection of the AI against premature degradation [24]. This not only increases pest control agent uptake and effectiveness but also significantly reduces pesticide usage. There is currently little knowledge on nano-pesticides compared to conventional pesticides [25]. Some environmentally friendly agricultural pest control solutions that pose little risk to humans and are beneficial for other organisms, such as microbes, have gained attraction [26]. As a result, bio-pesticides are more environmentally friendly than chemical pesticides and are becoming more popular. Presently, the bio-pesticides market is growing in the following areas: microbial pesticides derived from *Bacillus thuringiensis* (Bt), *B. subtilis* and *B. cereus*; and fungal pesticides such as *Beauveria bassiana* (Bals.), *Metarhizium anisopliae*, *Paecilomyces lilacinus* and *Verticillium*; and viral pesticides, such as nuclear polyhedrosis virus. Successive research in the area of nanomaterials has implicated different aspects of agrochemicals, such as the formulation of the pesticide, plant growth regulators, pesticide by product analysis, and the advancement of bio-pesticides. To enhance the yield in agriculture, pesticides are the most widely used agrochemicals [27].

The presence of nanomaterials increases the permeation, visibility and uptake of AIs on the target when pesticides are constructed using various strategies. This will help improve their utilization while also alleviating or reducing environmental pollution, contributing the sustainability of agricultural systems and improved food security [28]. In addition, in 2019, the International Union of Pure and Applied Chemistry identified ten chemical technologies that will have an impact on humans: nano-pesticides ranked first due to their potential reduced impact on the environment and public health [29].

Nanomaterials for pesticide formulation show a great effect in controlling biological disasters and increasing crop productivity. Solid formulations necessitate the use of solid emulsifiers and a variety of inert fillers, and they are widely used due to their ease of storage and transportation. During the manufacturing process dust particles are generated, and this can cause severe damage to health. Other organisms that come into contact are also affected. Hence it is necessary to maintain effective protection measures during manufacture [30, 31]. In conventional pesticides the main formulations are wettable powder and emulsifiable concentrate [32]. Nanomaterials used in pesticide production are primarily involved in the development of nano-pesticide formulation systems to improve the bioavailability and stability of AIs. AIs that are easily photolysed and decomposed can be stabilized under the safety of nanomaterials.

The other desirable compound is avermectin, which is susceptible to ultraviolet (UV) light. The stability of avermectin is improved using nanomaterials [33]. Nanomaterials can deliver AIs at the optimal working concentration in a targeted and controlled manner [34]. This helps to decrease the risk of resistance caused by the AI concentrations [35]. Also, it helps reduce waste of AIs and accidental damage to non-targets [36]. In situ deposition was used to create a magnetic nanocarrier made of diatomite and Fe_3O_4 [37]. Bioactivity studies revealed that DQ@MSN-SO$_3$ nanoparticles outperformed the control in terms of herbicidal activity against *Datura stramonium* L., using an attapulgite-based hydrogel [38].

With a core shell structure, a nanocomposite was prepared using palygorskite, glyphosate, and amino silicone oil. Using a magnetic field, the release of chemical glyphosate could be accelerated and the herbicides could be prevented from damaging the environment using an incomplete coating of amino silicon. Biosafety tests on zebrafish revealed that this delivery system was safer than glyphosate [39], because of their antibacterial or insecticidal properties [40]. Nanomaterials could be used directly as nano-pesticides. Nanomaterials easily penetrate cell membranes, due to their nanostructure and small size and the surface properties of the cell membrane; they affect normal cell organ function and immediately generate abnormal reactive oxygen species [41].

2.5   TYPES OF CONVENTIONAL PESTICIDES

According to the official website of the USEnvironmental Protection Agency, all ingredients, other than biological and antimicrobial pesticides, are called conventional pesticides. They are basically synthetic chemicals. Their work is to destroy or kill any pest that can also be just as beneficial for the plant growth, such as a desiccant or nitrogen stabilizer. Antifoulant and wood preservatives may also be considered conventional pesticides if they are not used as antimicrobials. Biochemical pesticides that are toxic are also considered as conventional pesticides. Biochemical pesticides or bio-pesticides are natural products used to protect the crops and kill pests. Biochemicals are generally of two types: microbial and biochemical. In microbial pesticides, living cells are used, such as bacteria, fungi, yeast and viruses, whereas in biochemical pesticides, plant extracts, semichemical and organic acids are used. Bio-pesticides are pesticides that are targeted to specific pests. Conventional pesticides are generally made by liquid emulsion, aerosol spray or water dispersion methods and there is a long range of products. Before these products can be used as pesticides, their chemistry, toxicity, performance on the active site, amount, and effect on pests are evaluated following the guidelines set out by the Federal Insecticides, Fungicide and Rodenticide Act. The efficacy data for all the pesticides are also investigated as regards the effectiveness of these conventional pesticides in target pests [42–44].

2.6   TYPES OF NANO-PESTICIDES

Studies have shown that nano-pesticides based on metals and agrochemicals are less toxic and are good alternatives to conventional pesticides. Essential oil- and bioactive

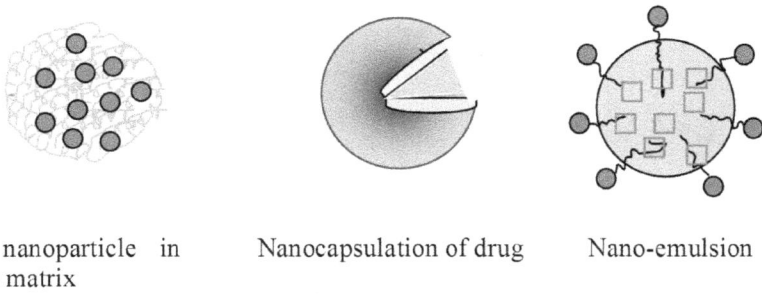

Metallic nanoparticle in   Nanocapsulation of drug   Nano-emulsion
polymeric matrix

FIGURE 2.1   Synthesis of nano-pesticides by various methods.

agent-based pesticides are also used in organic farming and sustainable agriculture [45]. The methods of synthesis of important nano-pesticides [45, 46] are depicted in Figure 2.1.

1. *Metals and their oxide nanoparticles*: Silver nanoparticles, zinc oxide nanoparticles, alumina dust, silicon oxide nanoparticles, titanium dioxide nanoparticles, copper nanoparticles
2. *Synthetic and natural polymer-based nanoparticles*: Polymer-based nano-pesticides, nanozeolites
3. *Nanoemulsions*: Nanoemulsions are used for storing grains. Nanoemulsions are very water-soluble and low-cost. The efficacy of these nanoemulsions can be increased on the basis of additives and cleanser ratio.

2.7 METHODS FOR SYNTHESIS OF NANO-PESTICIDES

2.7.1 EMULSION/NANOEMULSION

The mixture of oil and water makes an emulsion. When a nanoemulsion is being made, the size of the oil/water droplet becomes in the 100-nm to micrometre range. This size differentiates this emulsion from conventional emulsions. Because of the slow creaming rate, a nanoemulsion becomes more stable and homogeneous. Many nanoemulsions are on the market as micro-toxin inhibitors and antifungal for crop pests, e.g. citronella and neem oil nanoemulsion, *Pimpinella anisum* L. oil nanoemulsion, citral nanoemulsion with surfactants. All of these emulsions are good enough as nano-pesticides. Emulsions can be classified on the basis of stability, whether thermodynamically or kinetically stable. Nanoemulsions of permethrin and neem oil have been synthesized by Indian scientists, who found that the size of the emulsion depends on the activity of the nano-pesticide. Nanoemulsions of fatty acid methyl esters, organosilicones, alkyl glucosides and tebuconazole have also been synthesized and improved efficiency was found against different types of pests [46, 47] (Figure 2.2).

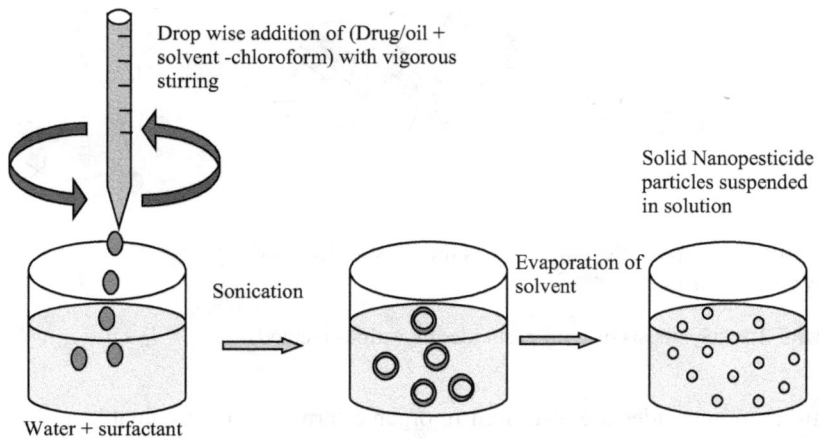

Drop wise addition of (Drug/oil + solvent -chloroform) with vigorous stirring

Solid Nanopesticide particles suspended in solution

Sonication

Evaporation of solvent

Water + surfactant

FIGURE 2.2   Synthesis of nano-pesticides by the nanoemulsion method.

2.7.2 Encapsulation

Encapsulation of plant extracts and different concentrations of essential oils that have different chemical ingredients has been found to be a very successful and effective method for nano-pesticide formation. Many scientists are working on the in-situ synthesis of nanoparticles and their encapsulation by biopolymers. Nano-pesticides based on copper using *Annona squamosa* L. seed have been synthesized and their efficacy on insect pests have been tested. It was concluded that copper nanoparticles had remarkable toxicity towards larvae of *Anopheles stephensi* and *Tenebrio molitor* [48]. Controlled-release nanocapsules where the Ais slowly degrade enhance the efficiency of pesticides [49].

2.7.3 Nanosuspension

Nanosuspension is dispersion of AIs in any liquid medium. This process is done using a surface-active agent. Surface-active agents or surfactants rearrange their molecules in the polar side of water/liquid and non-polar side in the pesticide nanoparticle.

2.7.4 Polymer-Based Nano-Pesticides

Different types of polymers, like guar gum, chitosan, starch, glycol and its polymer and polyesters have been used in the synthesis of polymer-based nano-pesticides [49–52]. These carbohydrate polymers have mostly been used as stabilizers and reducing agents for metal and metal oxide-based nano-pesticide formation. These biodegradable or biopolymers have been also used to cap AI. Polyethylene glycol capsulated temephos and imidacloprid nano-pesticides have been synthesized and found good against larvae of *Culex quinquefasciatus* [53]. Poly epsilon-caprolactone nanocapsules have been synthesized for controlled release of atrazine [50]. For *Blattella germanica*

TABLE 2.1
Metals and Metal Oxides as Nano-Pesticide

S. no.	Nanoparticle/nanoxide particle	Against	Controlled release	References
1	Silica nanoparticles coated with (3-mercaptopropyl) trimethoxysilane	Insects		[57]
2	Calcium carbonate and porous silica particles	Longer period of release	Validamycin	[58]
3	Cu nanoparticles	10,000 longer against bacteria for pomegranate		[59]
4	TiO_2 with Zn and silver nanodroplets	Bacterial spot disease in tomato and rose		[60, 61]
5	Alumina nanoparticles	Greater motility than *Sitophilus oryzae* and *Rhyzopertha dominica*		[62]

insect control, essential oil obtained from *Geranium* sp. and *Citrus reticulata* have been nanocapsulated with polyethylene glycol. In another study garlic essential oil coated with polyethylene glycol was also found to be useful in storing products and pest control [54].

2.7.5 NANOGELS

Pheromone-based nanogels have been synthesized using methyl eugenol to control fruit pests [55]. Chitosan-based nanogels with copper have been found to be very effective against *Fusarium graminearum* [56].

2.7.6 METALLIC NANOPARTICLE-BASED NANO-PESTICIDES

Metals and their oxides in nanoscale have shown very effective properties against pest, fungi and bacteria. Various mechanisms have been proposed by scientists for nano-pesticidal effects. Table 2.1 lists the metals and metal oxides used as nano-pesticides.

2.8 ADVANTAGES OF NANO-PESTICIDES OVER CONVENTIONAL PESTICIDES

Scientists estimate that 0.1% of pesticides reach their target by techniques like spray, soil and seed treatment. The remaining 99.9% of pesticides are wasted in ground water, soil and the environment. Non-selective pesticides also create a tough situation for naturally beneficial insects. The solubility of pesticides has always been a big challenge for scientists. Many pesticides are not soluble in water. So, a large number of organic solvents are required to dissolve these pesticides. This requires a good deal of manpower and chemicals, and increases the cost of crop production. Spray efficiency is another factor in pesticide efficacy. Pesticide efficacy depends on the stability of the AI in a pesticide. An ingredient that is mostly stable does not degrade in the environment; in rain it leaches into the soil, where it concentrates in plants or crops , before entering the animal and human cycle. This creates serious problems and even death. To mitigate these disadvantages, conventional pesticides are being replaced by nano-pesticides. Nano-scaling of pesticides decreases their non-target effects; a smaller amount of AIs is required to be used in pesticides, increasing efficiency. Nano-pesticide formation reduces the problem of the solubility of conventional pesticides. Using a smaller amount of AI reduces the cost of nano-pesticide as well as increasing crop production prices. The problem of solubility of a hydrophobic ingredient can easily be removed by nanocapsulation, nanogel, nanoemulsion or micelle formation methods.

S. no.	Pesticide properties	Nano-pesticide effect
1	Stability and solubility	⬆
2	Efficacy and uptake	⬆
3	Controlled release	⬆
4	Penetration and adhesive nature	⬆
5	Bioavailability	⬆
6	Environmental risk	⬇
7	Pest resistance	⬇
8	Active ingredient amount	⬇

Nano-pesticides also reduce the sharp smell of chemicals or ingredients used in pesticides, reduce evaporation and increase the activity of ingredients. There are many benefits of nano-pesticides over conventional pesticides; however, they are still toxic and hazardous for both humans and animals because of their greatly increased penetration properties. Nano-sizes also have specific properties that develop when conventional pesticides are converted to nano-pesticides.

2.9 CONCLUSION AND FUTURE PROSPECTS

Nano-pesticides are essential for best production, less toxicity, improved environmental hazardous effects, cost-effective, food security and essential nutrients for the crop. The good stability and controlled release of AIs in nano-pesticides increase the need for crop protection from pests. Improving the solubility problem of the AI in pesticides is an additional advantage for their use as nano-pesticides. Many more possibilities are still to be improved in nano-pesticides for the greatest possible environmentally friendly and financial approach. Some of the things that should be included are listed below:

1. Green synthesis of nanoparticles or nano-pesticide, whether metallic or polymeric, should be increasingly used.
2. The environmental effect of nano-pesticides should be considered in practical labs before they are rolled out on a large scale.
3. New ideas and types of eco-friendly methods and techniques for the synthesis of nano-pesticides should be encouraged.
4. Methods of nano-pesticide formation should be less time consuming and productive and effective on a large scale.
5. Nano-size also develops optical and visual properties: all toxic effects should be well researched before they are used in the field.
6. Enough medicinal plant extracts and their essential oil-based nano-pesticides, synthesized by the nanoemulsion method, may have been found for many pests and larvae. They are not material for humans and animals.
7. Nanoparticles that are synthesized from plant extract, biopolymer or green synthesis approaches should be increased.
8. A complete environmental study of nano-pesticides should be carried before they are used. Nanoparticles change their nature according to physical and chemical conditions. Silica nanoparticles have been considered non-toxic for plants, but some researchers have observed that the addition of silica nanoparticles decreases the pH of a medium, and behaves as a toxin in nature.

REFERENCES

1. Seshan K. Handbook of Thin Film Deposition: Techniques, Principles, Methods, Equipment and Applications. 2002; II Ed. CRC Press, Boca Raton, FL. 72p.
2. Ostiguy C, Roberge B, Woods C, Soucy B. Engineered nanoparticles: current knowledge about occupational health and safety risks and prevention measures. 2010; Studies and Research Projects/Report R-656, Montreal, IRSST.
3. Sebastian V, Arruebo M, Santamaria J. Reaction engineering strategies for the production of inorganic nanomaterials. Small, 2014; 10(5): 835–853.
4. Lövestam G, Rauscher H, Roebben G, Klüttgen BS, Gibson N, Putaud JP, Stamm H. Considerations on a definition of nanomaterial for regulatory purposes. Joint Research Centre (JRC) Reference Reports, 2020; 80: 00–41.
5. Vollath D. Nanomaterials an introduction to synthesis, properties and application. Environmental Engineering and Management Journal, 2008; 7(6): 865–870.
6. Benelmekki M. An introduction to nanoparticles and nanotechnology. In: Designing Hybrid Nanoparticles. 2015; Morgan & Claypool Publishers, San Rafael, CA. pp.1–14.
7. Khan MM, Kumar S, Alhazaa AN, Al-Gawati MA. Modifications in structural, morphological, optical and photocatalytic properties of ZnO: Mn nanoparticles by sol-gel protocol. Materials Science in Semiconductor Processing, 2018; 87: 134–141.
8. Buzea C, Pacheco II, Robbie K. Nanomaterials and nanoparticles: sources and toxicity. Biointerphases, 2007; 2(4): MR17–MR71.
9. Sun Y, Liang J, Tang L, Li H, Zhu Y, Jiang D, Song B, Chen M, Zeng G. Nano-pesticides: a great challenge for biodiversity? Nano Today, 2019; 28: 100757.

10. Klaine SJ, Alvarez PJ, Batley GE, Fernandes TF, Handy RD, Lyon DY, Mahendra S, McLaughlin MJ, Lead JR. Nanomaterials in the environment: behavior, fate, bioavailability, and effects. Environmental Toxicology and Chemistry: An International Journal, 2008; 27(9): 1825–1851.

11. Agathokleous E, Feng Z, Iavicoli I, Calabrese EJ. Nano-pesticides: a great challenge for biodiversity? The need for a broader perspective. Nano Today, 2020; 30: 100808.

12. Iavicoli I, Leso V, Beezhold DH, Shvedova AA. Nanotechnology in agriculture: opportunities, toxicological implications, and occupational risks. Toxicology and Applied Pharmacology, 2017; 329: 96–111.

13. Agathokleous E, Feng Z, Iavicoli I, Calabrese EJ. The two faces of nanomaterials: a quantification of hormesis in algae and plants. Environment International, 2019; 131: 105044.

14. Belz RG, Piepho HP. Predicting biphasic responses in binary mixtures: pelargonic acid versus glyphosate. Chemosphere, 2017; 178: 88–98.

15. Liu Y, Guo R, Tang S, Zhu F, Zhang S, Yan Z, Chen J. Single and mixture toxicities 380 of BDE-47, 6-OH-BDE-47 and 6-MeO-BDE-47 on the feeding activity of *Daphnia magna*: from behavior assessment to neurotoxicity. Chemosphere, 2018; 195: 542–550.

16. Agathokleous E, Feng Z, Iavicoli I, Calabrese EJ. The two faces of nanomaterials: a quantification of hormesis in algae and plants. Environment International, 2019; 31: 105044.

17. Chamsi O, Pinelli E, Faucon B, Perrault A, Lacroix L, Sánchez-Pérez JM, Charcosset JY. Effects of herbicide mixtures on freshwater microalgae with the potential effect of a safener. Annales de Limnologie-International Journal of Limnology, 2019; 55: 3.

18. Sharma VK, Sayes CM, Guo B, Pillai S, Parsons JG, Wang C, Yan B, Ma X. Interactions between silver nanoparticles and other metal nanoparticles under environmentally relevant conditions: a review. Science of the Total Environment, 2019; 653: 1042–1051.

19. Rather GA, Gul MZ, Riyaz M, Chakravorty A, Khan MH, Nanda A, Bhat MY. Toxicity and risk assessment of nanomaterials. *In*: Handbook of Research on Nano-Strategies for Combatting Antimicrobial Resistance and Cancer. 2021; IGI Global. Hershey, PA. pp. 391–416.

20. Petersen EJ et al. Determining what really counts: modeling and measuring nanoparticle number concentrations. Environmental Science: Nano, 2019; 6: 2876–2896.

21. Van Der Zande M, Kokalj AJ, Spurgeon DJ, Loureiro S, Silva PV, Khodaparast Z, Drobne D, Clark NJ, Van Den Brink NW, Baccaro M, Van Gestel CA. The gut barrier and the fate of engineered nanomaterials: a view from comparative physiology. Environmental Science: Nano, 2020; 7(7): 1874–1898.

22. Lead JR, Batley GE, Alvarez PJ, Croteau MN, Handy RD, McLaughlin MJ, Judy JD, Schirmer K. Nanomaterials in the environment: behavior, fate, bioavailability, and effects—an updated review. Environmental Toxicology and Chemistry, 2018; 37(8): 2029–2063.

23. Kah M. Nanopesticides and nano fertilizer: emerging contaminants or opportunities for risk mitigation? Frontiers in Chemistry, 2015; 3: 64.

24. Walker GW, Kookana RS, Smith NE, Kah M, Doolette CL, Reeves PT, Lovell W, Anderson DJ, Turney TW, Navarro DA. Ecological risk assessment of nano-enabled pesticides: a perspective on problem formulation. Journal of Agricultural and Food Chemistry, 2018; 66(26): 6480–6486.

25. Chaud M, Souto EB, Zielinska A, Severino P, Batain F, Oliveira-Junior J, Alves T. Nanopesticides in agriculture: benefits and challenge in agricultural productivity, toxicological risks to human health and environment. Toxics, 2021; 9(6): 131.

26. Tang FH, Lenzen M, McBratney A, Maggi F. Risk of pesticide pollution at the global scale. Nature Geoscience, 2021; 14(4): 206–210.
27. de Oliveira JL, Campos EVR, Bakshi M, Abhilash PC, Fraceto LF. Application of nanotechnology for the encapsulation of botanical insecticides for sustainable agriculture: prospects and promises. Biotechnology Advances, 2014; 32(8): 1550–1561.
28. Huang B, Chen F, Shen Y, Qian K, Wang Y, Sun C, Zhao X, Cui B, Gao F, Zeng Z, Cui H. Advances in targeted pesticides with environmentally responsive controlled release by nanotechnology. Nanomaterials, 2018; 8(2): 102.
29. Kumar V, Vaid K, Bansal SA, Kim KH. Nanomaterial-based immunosensors for ultrasensitive detection of pesticides/herbicides: current status and perspectives. Biosensors and Bioelectronics, 2020; 165: 112382.
30. Lade BD, Gogle DP. Nano-biopesticides: synthesis and applications in plant safety. *In*: Nanobiotechnology Applications in Plant Protection. 2019; Springer, Cham. pp. 169–189.
31. Mahmoudpour M, Karimzadeh Z, Ebrahimi G, Hasanzadeh M, Ezzati Nazhad Dolatabadi J. Synergizing functional nanomaterials with aptamers based on electrochemical strategies for pesticide detection: current status and perspectives. Critical Reviews in Analytical Chemistry, 2021; 1–28. https://doi.org/10.1080/10408 347.2021.1919987
32. Fincheira P, Tortella G, Seabra AB, Quiroz A, Diez MC, Rubilar O. Nanotechnology advances for sustainable agriculture: current knowledge and prospects in plant growth modulation and nutrition. Planta, 2021; 254(4): 1–25.
33. Dikbaş N, Cinisli KT. Microbial metabolites powered by nanoparticles could be used as pesticides in the future? Biotechnology Journal International, 2019; 23: 1–4.
34. Udomsap AD, Hallinger P. A bibliometric review of research on sustainable construction, 1994–2018. Journal of Cleaner Production, 2020; 254: 120073.
35. Kaur R, Mavi GK, Raghav S, Khan I. Pesticides classification and its impact on environment. International Journal of Current Microbiology and Applied Science, 2019; 8(3): 1889–1897.
36. An C, Cui J, Yu Q, Huang B, Li N, Jiang J, Shen Y, Wang C, Zhan S, Zhao X, Li X. Polylactic acid nanoparticles for co-delivery of dinotefuran and avermectin against pear tree pests with improved effective period and enhanced bioactivity. International Journal of Biological Macromolecules, 2022; 206: 633–641.
37. Kaziem AE, Gao Y, Zhang Y, Qin X, Xiao Y, Zhang Y, You H, Li J, He S. α-Amylase triggered carriers based on cyclodextrin anchored hollow mesoporous silica for enhancing insecticidal activity of avermectin against *Plutella xylostella*. Journal of Hazardous Materials, 2018; 359: 213–221.
38. Zhou M, Xiong Z, Yang D, Pang Y, Wang D, Qiu X. Preparation of slow release nanopesticide microspheres from benzoyl lignin. Holzforschung, 2018; 72(7): 599–607.
39. Liu B, Zhang J, Chen C, Wang D, Tian G, Zhang G, Cai D, Wu Z. Infrared-light-responsive controlled-release pesticide using hollow carbon microspheres@ polyethylene glycol/α-cyclodextrin gel. Journal of Agricultural and Food Chemistry, 2021; 69(25): 6981–6988.
40. Xiang Y, Zhang G, Chi Y, Cai D, Wu Z. Fabrication of a controllable nanopesticide system with magnetic collectability. Chemical Engineering Journal, 2017; 328: 320–330.
41. Wu F, Harper BJ, Crandon LE, Harper SL. Assessment of Cu and CuO nanoparticle ecological responses using laboratory small-scale microcosms. Environmental Science: Nano, 2020; 7(1): 105–115.

42. Ai L, Su J, Wang M, Jiang J. Bamboo-structured nitrogen-doped carbon nanotube encapsulating cobalt and molybdenum carbide nanoparticles: an efficient bifunctional electrocatalyst for overall water splitting. ACS Sustainable Chemistry & Engineering, 2018; 6(8): 9912–9920.

43. Kah M, Kookana RS, Gogos A et al. A critical evaluation of nanopesticides and nanofertilizers against their conventional analogues. Nature Nanotech, 2018; 13: 677–684.

44. Djiwanti SR, Kaushik S. Nanopesticide: future application of nanomaterials in plant protection. *In*: Prasad R (ed.) Plant Nanobionics. Nanotechnology in the Life Sciences. 2019; Springer, Cham. pp. 255–298.

45. Shekhar S, Sharma S, Kumar A, Taneja A, Sharma B. The framework of nanopesticides: a paradigm in biodiversity. Materials Advances, 2021; 2: 6569.

46. Camara MC, Campos EVR, Monteiro RA et al. Development of stimuli-responsive nano-based pesticides: emerging opportunities for agriculture. Journal of Nanobiotechnology, 2019; 17: 100.

47. Jasrotia P et al. Nanomaterials for postharvest management of insect pests: current state and future perspectives. Frontiers in Nanotechnology, 2022; 811056.

48. Rai M, Kon K, Ingle A, Duran N, Galdiero S, Galdiero M. Broad-spectrum bioactivities of silver nanoparticles: the emerging trends and future prospects. Applied Microbiology and Biotechnology, 2014; 98(5): 1951–1961.

49. Vivekanandhan P, Swathy K, Thomas A, Krutmuang P, Kweka EJ , Rahman A , Pittarate S , Krutmuang P. Insecticidal efficacy of microbial-mediated synthesized copper nano-pesticide against insect pests and non-target organisms. International Journal of Plant, Animal and Environmental Sciences, 2021; 11(3): 456.

50. Kashyap PL, Xiang X, Heiden P. Chitosan nanoparticle based delivery systems for sustainable agriculture. International Journal of Biological Macromolecules, 2015; 77: 36–51.

51. Hwang IC, Kim TH, Bang SH, Kwon KK, Seo HR et al. Effect of controlled release formulations of etofenprox based on nano-bio technique. Journal of the Faculty of Agriculture, Kyushu University, 2011; 56: 33–40.

52. Li M, Huang Q, Wu Y. A novel chitosan-poly (lactide) copolymer and its submicron particles as imidacloprid carriers. Pest Management Science, 2011; 67(7): 831–836.

53. Loha KM, Shakil NA, Kumar J, Singh MK, Srivastava C. Bio-efficacy evaluation of nanoformulation of β-cyfluthrin against *Callosobruchus maculatus* (Coleoptera: Bruchidae). Journal of Environmental Science and Health, 2012; 47(7): 687–691.

54. Oliveira JL, Campos EV, Gonçalves da Silva CM, Pasquoto T, Lima R et al. Solid lipid nanoparticles co-loaded with simazine and atrazine: preparation, characterization, and evaluation of herbicidal activity. Journal of Agricultural and Food Chemistry, 2015; 63(2): 422–432.

55. Yang FL, Li XG, Zhu F, Lei CL. Structural characterization of nanoparticles loaded with garlic essential oil and their insecticidal activity against *Tribolium castaneum* (Herbst) (Coleoptera: Tenebrionidae). Journal of Agricultural and Food Chemistry, 2009; 57(21): 10156–10162.

56. Jiang LC, Basri M, Omar D, Rahman MB, Salleh AB et al. Green nano-emulsion intervention for water-soluble glyphosate isopropylamine (IPA) formulations in controlling *Eleusine indica* (*E. indica*). Pesticide Biochemistry and Physiology, 2012; 102(1): 19–29.

57. Lim CJ, Basri M, Omar D, Rahman MB, Salleh AB et al. Physicochemical characterization and formation of glyphosate-laden nano-emulsion for herbicide formulation. Industrial Crops and Products, 2012; 36(1): 607–613.
58. Bhagat D, Samanta K, Bhattacharya S. Efficient management of fruit pests by pheromone nanogels. Scientific Reports, 2013; 3: 1294.
59. Brunel F, El Gueddari NE, Moerschbacher BM. Complexation of copper (II) with chitosan nanogels. Toward control of microbial growth. Carbohydrate Polymers, 2013; 92(2): 1348–1356.
60. Paret ML, Vallad GE, Averett DR, Jones JB, Olson SM. Photocatalysis: effect of light-activated nanoscale formulations of TiO$_2$ on *Xanthomonas perforans* and control of bacterial spot of tomato. Phytopathology, 2013; 103(3): 228–236.
61. Ocsoy I, Paret ML, Ocsoy MA, Kunwar S, Chen T et al. Nanotechnology in plant disease management: DNA-directed silver nanoparticles on graphene oxide as an antibacterial against *Xanthomonas perforans*. ACS Nano, 2013; 7(10): 8972–8980.
62. Buteler M, Sofie SW, Weaver DK, Driscoll D, Muretta J et al. Development of nanoalumina dust as insecticide against Sitophilusoryzae and Rhyzoperthadominica. International Journal of Pest Management, 2015; 61(1): 80–89.

3 Nanomaterials as Nano-Fertilizers

Efat Zohra, Muhammad Ikram, Naveed Iqbal Raja, Zia-Ur-Rahman Mashwani, Ahmad A. Omar, Azza H. Mohamed, Seyed Morteza Zahedi, and Azhar Abbas

CONTENTS

3.1 INTRODUCTION

Nanomaterials possess unique features and their properties differ significantly from bulk material. This is mainly due to their nano-size which renders them high surface energy, large surface area, and spatial confinement. These properties do not usually occur in a bulk material. The size and surface area of any material determine how it interacts with any biological system. The decrease in size of the material increases the surface area and enhances its ability to react with any surface with which it comes into contact [1]. It has been observed that the efficiency of various biological systems depends on the size of the material. Because of the above-mentioned properties, the nanomaterial is currently employed in heavy-metal remediation, electronics, gene therapy, drug delivery, tissue culture engineering, crop bio-fortification, and crop protection [2, 3]. Nanobiotechnology provides a promising contribution to agriculture which could lead to new ways to overcome the various problems of the agriculture sector. Keeping in mind the magnificent properties of nanomaterials, plant researchers are using them as nano-fertilizers for crop protection and production. Green nano-technology has revolutionized the agriculture sector all around the world. Meanwhile, the different concentrations of nanomaterial-based nano-fertilizers have been having

DOI: 10.1201/9781003322122-3

promising effects on the amelioration of nutritional values of crops, ensuring fewer economic losses [4]. Nano-fertilizers are modified versions of conventional fertilizers in which plant micronutrients are encapsulated into nanoparticles (NPs) and delivered in the form of nano-sized emulsions. Furthermore, nanomaterials have promising application potential in the agriculture field, i.e. plant pest and pathogen control, nano-fertilizers, controlled release of insecticides and pesticides, efficient nutrient utilization, and soil fertility [5, 6]. In past decades, the increasing population has forced more production of food to fulfill the needs of millions of people in developing and non-developing countries. Unfortunately, nutrient deficiency in the soil could lead to qualitative and quantitative nutrient imbalance for both humans and livestock, creating economic losses in agro-business [7]. Additionally, uptake and utilization of essential nutrients are low because of their adhesion with soil particles. Generally, only a small amount of nutrients is present in chemical-based fertilizers efficiently used by plants. In conclusion, conventional or chemical-assisted fertilizers may not the most suitable option due to their reduced bio-availability, bio-degradability, bio-compatibility, and potential environmental risks [4].

Keeping in mind the points discussed above, it is an inevitable need to synthesize nano-based material that has the properties of efficient controlled release of nutrients to targeted sites in plants to overcome the problems of agricultural crops. In this scenario, various selenium (Se)-, silver (Ag)-, iron (Fe)-, titanium (Ti)-, and copper (Cu)-based nanomaterials have been employed for the improvement of plants subjected to biotic and abiotic stresses [8–10]. These nanomaterials can enhance crop productivity by enhancing the uptake and utilization of fertilizers by the plants. Additionally, it has been reported in many studies that these nanomaterials can mitigate crop diseases by killing a variety of plant pathogens, including the generation of reactive oxygen species [11]. Moreover, the use of nano-fertilizers may also overcome the deleterious environmental impacts of traditional chemical-based fertilizers. Besides this, some studies have also revealed that nano-fertilizers promisingly improve crop productivity by increasing the seed germination percentage, seedling growth, and photosynthetic activity. These nano-fertilizers are also involved in activating the antioxidant defense system of plants [12, 13].

Although nanomaterials are at an early stage of development, the ethical and safety concerns associated with their use in crop production have been relatively unexplored. Therefore, such nano-fertilizers must be carefully evaluated before being made available for use. In this chapter, we highlight the potential applications and mechanistic routes that have been adopted by various nano-fertilizers for the betterment of crops subjected to biotic and abiotic stresses.

3.2 CONVENTIONAL AND BIOINSPIRED FERTILIZERS

To improve crop production and its quality, fertilization is necessary for soil fertility. Precision nutrient management is an emerging challenge around the world because it is entirely dependent on conventional or chemical fertilizers. Further, chemical-based fertilizers are considered to be effective for enhancing crop production [14]. Unfortunately, less than half the amount of applied chemical-based fertilizers uptake by the plants and the remaining amount of nutrients may be leached out and become fixed in soil, ultimately leading to possible soil and water pollution [15]. According to different studies, important nutrients, such as nitrogen, potassium, and phosphorus, are lost by 40–70%, 80–90%, and 50–90%, respectively, resulting in a huge loss of resources [16]. In addition, the extensive use of chemical-based fertilizers by farmers to attain desired maximum production could lead to an increase in the accumulation of salts which reduce soil fertility, ultimately resulting in crop losses. Furthermore, overuse of fertilizers without their controlled release can have deleterious effects on the quality of crops. Besides this, chemical fertilizers are not cost-effective for farmers and also may have environmental toxicity issues [15]. Hence there is an urgent need to search for and formulate environmentally friendly fertilizers that are limited to the target site to enhance the quality and quantity of crops. In this scenario, bioinspired nano-fertilizers offer various benefits in nutrition management because of their excellent nutrient use efficiency. This management strategy not only enhances nutrients but also reduces nutrient leaching into the water. Further, biogenic nano-fertilizers are also used to enhance biotic and abiotic stress tolerance in crops. Many studies have reported that various biological resources have been employed for the synthesis of nanomaterial, so-called nano-fertilizers [17]. Plant extract-mediated nanomaterials are eco-friendly, cost-effective, target-specific, and biocompatible because of the extra coating of phytochemicals on their surfaces. Various kinds of nanomaterials, such as SeNPs, CuNPs, ZnNPs, FeNPs, and TiO_2NPs, are commonly used in agriculture to enhance crop production. To meet the nutrient requirements of plants, nano-fertilizers contain nutrients encapsulated in NPs for controlled release [18]. Surprisingly, these nano-fertilizers are commonly regarded as the best alternative approach to enhance crop quality and quantity as compared to traditional chemical-based fertilizers. The nano-fertilizers are involved in the effective absorption of nutrients and vital components for the plants. Nano-fertilizer exposure in agriculture is helpful to achieve sustainable crop production. Bioinspired nano-fertilizers have important advantages over traditional chemical-based fertilizers [19]. Furthermore, nano-fertilizers are cost-effective and reduce the need for transportation and its cost. It is reported that nano-fertilizers allow farmers to provide specific amounts of nutrients to a particular crop. The plants also benefit from bioinspired nano-fertilizers in overcoming biotic and abiotic stresses [20]. The main attention on the use of bioinspired nano-fertilizers is on the role of various natural phytochemicals which are involved in the reduction, reduction, capping, and stabilization of nanomaterials [21].

3.3 NANOTECHNOLOGY IN AGRICULTURE

A major sector of the economy is agriculture as it provides raw materials for the food and feed industries. Agriculture is the sole industry that produces food from both transitional and final inputs using well-known technologies. This makes it important to use innovative methods for agriculture. The development of agriculture in developing countries is relatively advantageous; however, the importance of food products still remains low [22, 23]. Agriculture also plays a pivotal role in the economy of developing countries. In the recent past, climate change has exerted a deleterious impact on crop production worldwide. Climate changes have already affected the production of important crops, such as wheat, barley, and rice [24]. The major environmental stresses that exert pressure on the crops are increasing heat stress in arable areas, drought, salinity, accumulation of heavy metals, and various fungal and bacterial diseases. Furthermore, these devastating plant stresses may have a significant impact on crop yield and quality. The increase in environmental stresses has detrimental effects on crop yields and causes the accumulation of heavy metals in plant tissues, which makes plants unsuitable for livestock and humans and creates severe health issues [3].

Global food supply has been greatly increased as a result of agricultural practices associated with the green revolution. These agricultural practices also have a deleterious impact on the environment and agriculture ecosystem services, highlighting the inevitable need for more bio-compatible, cost-effective, and sustainable agricultural approaches. It is well documented and explored that inappropriate and excessive use of chemical-based fertilizers, insecticides, and pesticides has enhanced toxins and nutrients in groundwater, causing health-associated problems, water purification costs, and recreational opportunities [25, 26]. Conventional agricultural practices contribute to the killing of vital insects and other wildlife. As an ebullient sphere of science, nanotechnology has emerged as a viable solution for overcoming the risks associated with conventional farming practices in the agricultural sector. By definition, nanotechnology aims to understand and control matter at scales of up to 100 nm, where its unique physical properties allow for novel applications. Furthermore, nanotechnology could also be described as the science of designing and creating machines where every atom and a chemical bond are specifically tailored [27]. Nanotechnology has promising applications in the agriculture sector, including nano-pesticides, nano-fertilizers, and nano-insecticide formulation, to enhance nutrient levels to increase yield and productivity without soil and water contamination and also to provide protection against microbial diseases [28]. It is reported that nanotechnology acts as biosensors in agricultural fields to maintain the health of crops by monitoring soil nutrient quality. In this scenario, researchers have explored the applications of nanotechnology to improve crops by enhancing food security and crop sustainability. Nanotechnology also plays a role in crop production by controlling the release of nutrients [29]. Furthermore, nanotechnology is being used to monitor water quality, insecticides, and pesticides in order to ensure sustainable agriculture.

The application of nanotechnology in the agriculture sector is vital for sustainable agricultural development. Potential applications of nanotechnology, such as nano-filtration, biosensors, and controlled delivery systems, were reported in agro-food areas [30]. Furthermore, nanotechnology has been proved to be good in agriculture resource management, efficient drug delivery in plants, and maintaining soil fertility. In addition, nano-sensors are widely used in agriculture because of their specific characteristics in monitoring heavy metals and other contaminated agents in soil and water [31]. In recent decades, nano-fertilizers have been available in the market and these contain nano selenium, iron, silver, titanium, copper, etc. These nanostructures have attracted the attention of researchers in biological studies and their potential in agriculture because of their bio-compatibility, reduced toxicity, and unique optical properties [32]. Additionally, the use of nanostructures in crop protection and production of food have been less explored. It is well documented that insects and pests are affected by the crops but nanomaterials have a key role to control insects, pests, and host pathogens [33]. It is reported that nano-encapsulated pesticides have been developed with slow-release properties, specificity, permeability, stability, and enhanced solubility [34]. It is well documented that various NPs such as selenium, silver, iron, copper, and zinc NPs, have outstanding applications as nano-fertilizers to enhance agronomic, physiological, and biochemical profiling of crops, ultimately resulting in high productivity and yield [35, 36]. In conclusion, despite a lot of information about NPs, nanostructures and nanocomposite formulations are available but, unfortunately, the toxicological impacts of NPs are still not well defined, hence the application of nanomaterials is vast because of the lack of toxicity assessments and the human health-related risks. Therefore, there is a need for the development of a comprehensive database on nanotechnology and a well-organized alarm system, requiring international collaboration and proper legislation.

3.4   POTENTIAL APPLICATIONS OF NANOPARTICLES AS NANO-FERTILIZERS

The potential applications of nanomaterials as nano-fertilizers are given below.

3.4.1   SELENIUM NANOPARTICLES AS NANO-FERTILIZERS

Selenium is an essential trace element required for the normal physiological and biochemical functioning of plants and animals. Selenium is also a component of selenoproteins, which are antioxidant enzymes necessary for plant redox homeostasis, along with another antioxidant enzyme. Selenium acts as a cofactor for many antioxidative enzymes, including glutathione peroxidase, superoxide dismutase, and thioredoxin reductase, which help the body remove free radicals [2, 37]. Further, selenium is found in water, soil, animals, and crops. Selenium contents vary greatly throughout the world – from 0.005 to 1200 µg g^{-1} and most commonly between 0.1 and 10 µg g^{-1}. Recently, it has been reported that selenium has been used in agriculture for the bio-fortification of crops. Selenium has been found to overcome the deleterious effects of environmental stresses like drought, salinity, heat stress, and bacterial and fungal diseases. Selenium acts as an antioxidant at low concentrations but it acts as a pro-oxidant at high concentrations. It has also been demonstrated that selenium helps to maintain the structural integrity of the chloroplast and plastid membrane [3, 38]. Moreover, the appropriate dose of selenium can cause a decrease in electrolyte leakage, and an improved membrane stability index ultimately improved cellular integrity. Selenium can significantly enhance photosynthesis, delay leaf senescence, and increase the quality and quantity of crops [39, 40]. However, it is reported in many studies that overdose of selenium is involved in the overproduction of reactive oxygen species and damages plant bodies. In many studies, it has been revealed that selenium at a higher dose causes a significant decrease in leaf area and plant biomass and also disrupted the structural organization of plant cells [41]. Selenium NPs emerge as a bio-compatible nano-fertilizer to overcome the high dosage of selenium. In this scenario, selenium NPs have gained the attention of researchers, plant physiologists, and plant pathologists because of their remarkable potential applications in the agriculture sector. It has been reported in many studies that selenium NPs have been used as nano-fertilizers to enhance crop production under biotic and abiotic stresses [42, 43]. It has also been revealed that foliar applications of plant-based selenium NPs enhance drought tolerance in wheat plants. Garlic extract-mediated selenium NPs at 30 mg L^{-1} concentration fortify wheat crops under drought stress and enhance agronomic parameters such as root length, shoot length, plant biomass, shoot length, plant fresh and dry weight, leaf number, and leaf area of wheat plants [9].

Other researchers have also proposed that foliar spray of selenium NPs is a useful strategy for ameliorating strawberry plants and enhancing yield under salinity stress. They concluded that, under salinity stress, foliar applications of selenium NPs improve strawberry plant growth and yield parameters. Selenium NPs are involved in the protection of photosynthetic machinery, accumulation of total carbohydrates, and proline contents for osmo-protection. They also reported that selenium NPs alleviate salt stress by activating the activities of antioxidative enzymes for efficient reactive oxygen species homeostasis [44]. In addition, selenium NPs are well explored for the mitigation of heat stress in grain sorghum plants. During the booting stage of sorghum, foliar applications significantly stimulated the antioxidant defense system by up-regulating the activities of different enzymes involved in the antioxidant defense system. Selenium NPs also facilitated higher levels of unsaturated phospholipids. Selenium NPs also improved seed germination percentage, increase seed set percentage, reduced oxidative damage, and increased synthetic rate, ultimately enhancing the yield of crops [45, 46].

In addition, selenium NPs are very effective against bacterial and fungal diseases in plants. It has been demonstrated that plant extract-mediated selenium NPs at a concentration of 75 mg L^{-1} are more effective in treating huanglongbing disease in citrus plants by improving the photosynthetic capacity, activating enzymatic and non-enzymatic activities of plants. Researchers revealed that selenium NPs inhibit the growth of huanglongbing casing bacteria by breaking the cell wall, then binding with the cell membrane and altering the translation process and food metabolism, eventually leading to cell death [37].

In conclusion, selenium NPs as nano-fertilizers have the potential to increase nutrient uptake efficiency due to higher nutrient uptake caused by the small surface area of NPs which enhances nutrient–surface interaction. In addition, selenium NPs as nano-fertilizers not only enhance crop production under environmental stresses but also reduce the environmental pollution caused by conventional fertilizers. However, intensive field research experiments are required to explore the effectiveness of selenium NPs prior to their large-scale field application.

3.4.2 SILVER NANOPARTICLES AS NANO-FERTILIZERS

Silver NPs ($AgNO_3$) are the widely used NPs for the bio-system. It is reported that silver NPs have significant bactericidal and inhibitory effects as well as extensive spectrum antimicrobial applications. Nano-silver with a high surface area and a high fraction of surface atoms has remarkable antimicrobial potential compared to bulk silver. Further, silver NPs are well reported for their potential outstanding antifungal, antibacterial, and antiviral activities [47, 48]. In the recent past, silver NPs have emerged as nano-fertilizers for crop protection and production under biotic and abiotic stresses. It is reported that biosynthesized silver NPs have been used as a strong potential agent to enhance plant growth and seed germination percentage. However, silver NPs showed a dose-dependent response in plants with beneficial and negative effects. It is reported that plant-based nano-silver significantly influenced the agronomic parameters of wheat plants under heat stress [10].

Silver NPs are involved in interfering with the photosynthetic process and boosting the antioxidant defense system of plants, ultimately improving plant health. As a result of high heat stress, wheat plants are adversely affected in terms of agronomic parameters; however, the application of plant-based silver NPs protects the plants from a number of adverse effects and can yield better results in terms of shoot length (5 and 5.4%), fresh weight (1.3 and 2%), root number (6.6 and 7.5%), and dry weight (0.36 and 0.60%) than control plants [10]. Similarly, in another study, it was reported that various concentrations of silver NPs showed maximum seed germination, speed of germination, root length, root dry weight, and root fresh weight. In addition, silver NPs at a concentration of 50 ppm showed positive effects on root fresh weight and shoot length in *Brassica juncea* [49]. Furthermore, it was also observed that silver NPs have positive effects on brassica and cowpea plants. It was revealed that a 50 ppm concentration of silver NPs resulted in increased root nodulation, agronomic parameters like seed germination percentage, seed germination index, and seedling fresh and dry weight in cowpea plants [49].

Crops are facing the challenges of various pests and insects devastating the quality and quantity of crops. In this scenario, silver NPs are extensively in use for pest protection and bio-fortification of crops. It was reported that silver NPs destroy harmful microorganisms in the soil and limit the extensive use of conventional fertilizers in farming. It was demonstrated that silver NPs are responsible for preventing fungal and bacterial diseases in plants [50]. The foliar applications of various concentrations of silver NPs (10, 20, 30, 40 ppm) are very effective for controlling citrus canker diseases and brown spot disease. It is also reported that silver NPs enhance the antioxidant defense system of plants by lowering oxidative damage. Silver NPs significantly ameliorated the enzymatic and no-enzymatic activities in citrus plants suffering from biotic stress [51]. The exact antimicrobial mechanistic action of silver NPs is still unknown. However, some studies reported that the positive charge on silver NPs provides silver with the opportunity to combine with negative charges of the cell membrane and nucleic acids, resulting in destabilization of the plasma membrane, and it is also reported that silver NPs release reactive oxygen species that are involved in the breakdown of DNA structure which showed the antimicrobial potential of silver NPs [21]. In addition, it is also reported that silver NPs are more adhesive on the fungal and bacterial cell surface and showed remarkable fungicidal and bactericidal effects. Researchers formulated silver NPs using various plant extracts which showed outstanding antimicrobial potential against various crop pathogens. Plant-based silver NPs have shown positive effects on rice plants suffering from *Aspergillus flavus*. Silver NPs positively ameliorated physicochemical, biochemical, and agronomic parameters of rice plants under biotic stress. Further, silver NPs displayed a promising potential for the regulation of osmoprotectants, antioxidant enzymes (superoxide dismutase, peroxidase, catalase), and non-enzymatic (total flavonoid content and total phenolic content) metabolites and total protein contents in rice plants under fungal stress [52, 53]. The notable antimicrobial potential may have been attributed to the interaction of silver ions with various essential enzymes, and cellular proteins and the production of reactive oxygen species resulting in the inhibition of metabolic processes of the cell. Furthermore, plant-based silver NPs cause DNA damage and hamper cell division [54].

In conclusion, silver NPs are recognized as strong potential agents in agribusiness because of their magnificent antimicrobial and antioxidant potential applications. Silver NPs can be successfully used as nano-fertilizers to enhance crop protection from devastating fungal and bacterial diseases. Further, silver NPs have tremendous application potential as a new management strategic tool to resolve unsolved problems in the agriculture field.It is recommended that plant-based silver NPs may be an effective management strategy for agricultural problems because of their biocompatibility, bio-availability, eco-friendly nature, and cost-effective properties [55]. Now, there is a need for more research experiments to assess the impact of silver NPs on the environment and human health before their use on a large scale.

3.4.3 Titanium Dioxide Nanoparticles as a Nano-Fertilizer

Titanium dioxide NPs (TiO_2 NPs) are the emerging NPs that have significant applications as a nano-fertilizer in agriculture. Currently, many studies have revealed the importance of titanium dioxide NPs as the alternative to conventional chemical-based fertilizers. Titanium dioxide NPs are very useful to cope with environmental stresses. In the recent past, it was reported that plant-based titanium dioxide NPs have positive effects on the physiological parameters of wheat plants suffering from drought stress [56]. In this study, titanium dioxide NPs enhanced the morpho-physiological attributes of wheat plants under drought stress. The foliar spray of various concentrations of biogenic titanium dioxide NPs significantly ameliorated shoot length (53%) and shoot fresh (48%) and dry weight (44%) as compared to control plants. Similarly, a positive dose-dependent response of titanium dioxide NPs was observed in wheat plants for improving the level of antioxidant enzymes catalase (67%), superoxide dismutase (81%), and peroxidase (52%) as compared to control plants under drought stress [57]. They also reported that plant-based titanium dioxide NPs showed outstanding results at a concentration of 40 mg L^{-1},which improved the wheat plants subjected to drought stress.

Furthermore, another study displayed the beneficial effects of titanium dioxide on morphological traits. It was reported that foliar spray of 100 mg L^{-1} titanium dioxide NPs improved agronomic attributes and antioxidant defensive enzyme activity of Moldavian balm plants facing different levels of salinity stress. Titanium dioxide NPs also reduced the contents of H_2O_2 in Moldavian balm plants under salinity stress as compared to untreated plants [56]. Currently, various biotic stresses like bacterial and fungal diseases cause a huge loss to crops. To solve this problem, titanium dioxide NPs emerge as a magnificent strategy that has notable antimicrobial properties to combat plant diseases. It is reported that plant extract-assisted titanium dioxide NPs have the potential to treat fungal diseases in wheat plants. Researchers reported that titanium dioxide NPs increased the root length, shoot length, leaf area, leaf number, plant fresh and dry weight under *Bipolaris sorokiniana* stress. The researchers also reported that

40 mg L^{-1} concentration of titanium dioxide NPs influenced relative water content, chlorophyll content, total soluble sugars, phenolic and flavonoid contents of wheat plants under biotic fungal stress [11]. The mechanism by which titanium dioxide NPs work is to activate the antioxidant defense system in plants. These NPs mitigate the deleterious effects of overproduction of reactive oxygen species in plants subjected to environmental stresses [58]. Further, it is also revealed that plant-based NPs have various phytochemicals on their surface which boost the defense system, ultimately creating microbial resistance in the plants. Furthermore, titanium dioxide NPs promoted the activity of the rubisco enzyme, thus accelerating the agronomical traits of plants. Titanium dioxide NPs also improve nitrogen metabolism and convert inorganic nitrogen into organic nitrogen, ultimately enhancing plant growth. Interestingly, it was also reported that titanium dioxide NPs improve light-harvesting complex II contents and promoted antioxidant stress by lowering the accumulation of superoxide radicals and malondialdehyde content, and enhancing the activity of guaiacol peroxidase, ascorbate peroxidase, and catalase enzymes [59].

Keeping in view the immense potential application of titanium dioxide NPs in agriculture, it can be an excellent tool to mitigate the adverse effects of environmental stresses. In conclusion, titanium dioxide NPs as nano-fertilizers are extensively used to increase crop quality and quantity. In the near future, these NP-based nano fertilizers can be an outstanding solution for unsolved problems of agriculture. However, there are some unexplored NPs associated with potential risks which should be addressed. So there is also an inevitable need to determine the mechanism of action of NPs as nano-fertilizers.

3.4.4 IRON NANOPARTICLES AS NANO-FERTILIZERS

Iron (Fe) is a vital nutrient for plants and animals. It is reported that Iron deficiency is common among crops. Further, some studies have revealed that iron is high in soil but most of its proportion is usually fixed with the soil particles. Iron is involved in many physiological functions, including the synthesis of chlorophyll redox reaction and respiration [60]. In addition, iron deficiency significantly affects the growth and development of plants and causes anemia in animals and humans. Interestingly, iron is used as an active site for many vital redox enzymes dealing with the oxidation reduction reaction and cellular respiration in plants and animals [61]. Besides this, lack of iron could lead to cause chlorosis and a significant reduction in the photosynthesis phenomenon ultimately results in tree death. Therefore, it is an inevitable need to develop iron-based nano-fertilizers to improve crop production. In this scenario, nanomaterials are commonly used practically in agriculture sectors for the improvement of crop protection and production. According to some studies, iron oxide NPs could be used to replace traditional iron fertilizers. Researchers reported that iron oxide NPs have positive effects on the growth and development of peanuts (*Arachis hypogea*). The results revealed that iron oxide NPs enhance the developmental growth of peanuts by regulating the activity of antioxidant defensive enzymes and phytohormone activity [61]. It was also demonstrated that foliar applications of micronutrient oxide NPs of iron, manganese, and zinc significantly improve the growth quality and yield of squash plants. The results obtained from the experiment revealed that these NPs positively ameliorate the photosynthetic activity, organic matter, lipids, proteins, and fruit quality in squash plants [62].

Another study also displayed the positive response of green synthesized iron oxide NPs as nano-fertilizers for the alleviation of drought stress in plants. The results showed that iron oxide NPs can overcome the drastic effects of drought stress by enhancing the activity of photosynthetic mechanisms and also improving the soluble sugar content in *Setaria italica* subjected to drought stress [63]. According to their observation, iron oxide NPs have no toxicological impact on plants under drought stress, which confirms the eco-friendly nature of green synthesized iron oxide NPs as nano-fertilizers for agriculture crops. In addition, iron NPs have the application potential to overcome the deleterious effects of salinity stress in grape softwood cuttings of Khoshnaw cultivar. Experimental results revealed that iron NPs significantly enhanced total protein content and the antioxidant defense system and reduced the proline and hydrogen oxide content in grape softwood cuttings of Khoshnaw cultivar under salinity stress [64].

Besides this, iron oxide NPs have the potential to increase the leaf's fresh and dry weight, potassium, iron, phosphorus, and calcium content in peppermint under salinity stress. It is also reported that iron oxide NPs reduce the proline and peroxidation contents of peppermint under salinity stress. Experimental results proved that iron NPs can be a potential candidate as nano-fertilizer to overcome the effects of salinity in plants. These nanoparticles can also be used in agriculture as a nano-fertilizer instead of conventional chemical-based fertilizers [65].

3.4.5 COPPER NANOPARTICLES AS NANO-FERTILIZERS

Copper (Cu) is an essential micronutrient present in high concentrations in chloroplasts. It is reported that approximately 70% of total copper is present in the chloroplast. Interestingly, copper plays an important role in the biosynthesis of chlorophyll and is also involved in carbohydrate and protein metabolisms. In addition, copper is incorporated in metalloenzymes and many proteins and plays a significant role in plant health [66]. Generally, copper deficiency could cause various maladies, including stunted growth, distortion of young leaves, bleaching, and necrosis in plants. Further, its deficiency affects the formation of grains and fruits. Besides this, Cu deficiency also reduced cell wall lignification, leading to distortion of leaves, twisting, and bending of twigs and stems [67, 68]. In this scenario, copper NPs emerge as potential nano-fertilizers to solve the above-mentioned problems related to Cu-deficiency plants. Preliminary studies revealed that CuNPs show dose-dependent positive responses on the agronomic attributes and seed germination percentages in chickpea and soybean. The results obtained showed that CuNPs significantly ameliorated seed germination at a concentration of 2000 ppm. Furthermore, experimental results of other studies showed that CuNPs enhanced the seed germination percentage by 65% in soybean as compared to untreated control plants [69]. Similarly, it is reported that various concentrations (10, 20, 30, 40, 50 ppm) enhance the growth, biomass, and productivity of the wheat plant as compared to the control. In addition, it was demonstrated that a 30 ppm concentration of CuNPs enhanced chlorophyll contents, number of spikes, number of grains, and grain yields in wheat plants [70]. CuNPs also have the potential to enhance the bioactive content of tomato plants. Some published research studies have reported that various concentrations of CuNPs (50, 125, 250, 500 mg L^{-1}) have positive effects on the fruit quality and antioxidant activity of potato plants. According to the results obtained, CuNPs increase fruit firmness and enhance vitamin C, lycopene, and ABTS antioxidant capacity as compared to untreated plants. Additionally, plants treated with CuNPs showed increased levels of superoxide dismutase and catalase enzymes [71]. As a result, CuNP applications improve the quality of tomato fruits for human consumption due to the accumulation of bioactive compounds. Nano-formulations have application potential to treat microbial diseases in plants. Further, the use of nanomaterial as nano-fertilizer has reduced the use of agrochemicals, and petrochemicals are effective in the delivery of herbicides, fungicides, and pesticides. Therefore, in the recent past, nanotechnology in agriculture has emerged as having a great impact on plant protection. In this scenario, copper NPs are very effective against those bacteria and fungi causing various diseases in crops. Furthermore, some studies have reported that CuNPs are effective against bacterial blight in *Punica granatum* caused by bacteria *Xanthomonas axonopodis* [72]. It has also been reported that the green and chemically synthesized CuNPs have an inhibitory effect on the growth and development of *Fusarium graminearum*, *F. equiseti*, and *F. oxysporum* [73]. In conclusion, CuNPs will be the most sought-after nanomaterial in agriculture which could be used in plant protection.

3.5 LIMITATIONS OF NANO-FERTILIZERS AND FUTURE CHALLENGES

Nano-fertilizers are very important for sustainable agriculture to achieve crop protection and production under various biotic and abiotic environmental circumstances. However, extensive use of nano-fertilizers in agriculture sectors could have irreversible consequences. By keeping in mind environmental and health safety issues, it may be possible to reduce the use of nanotechnology products for crop enhancement. There is evidence from many published studies that plants respond differently to NPs based on their dose. Some researchers have explored the dose-dependent response of plant extract-mediated silver NPs in wheat plants under heat stress. They reported that the physiological performance of wheat plants was excellent at a concentration of 25–50 ppm but tended to decrease as the concentration was increased to 75–100 ppm [74]. Similarly, it has been reported that various concentrations of silver NPs (25, 50, 75, and 100 mg L^{-1}) were exogenously applied to wheat plants at the trifoliate stage. The researchers reported that silver NPs enhance the agronomic parameters of wheat plants but significant results were found at concentrations of 75 and 100 mg L^{-1} [10]. However, it is still necessary to investigate the optimum dose, suitable mode of application of NPs, and duration of NP treatment in mitigating the deleterious effects of heat stress on wheat plants in field conditions.

Other studies have also reported the dose-dependent response of selenium NPs. The results obtained showed that plant-based selenium NPs are very effective in controlling the huanglongbing disease in citrus plants by ameliorating their physicochemical, biochemical, and antioxidant defense system. The researchers reported that selenium NPs ameliorated the citrus physiology and biochemistry at minimal dosage. However, selenium NPs at a concentration of 100 mg L^{-1} decreased the physiological and biochemical attributes of citrus plants [37]. Similarly, in some studies it was reported that a low dose of plant-based titanium oxide NPs is very effective in controlling fungal diseases in wheat plants, although these showed deleterious effects on wheat plants as the concentration tended to increase at 60 and 80 mg L^{-1} [11]. By keeping in mind the above-mentioned concern of nanomaterials as a nano-fertilizer, it is important to consider the magnificent role of nano-fertilizers in agriculture, although it is also crucial to explore their drawbacks and side effects before market implementation. Nanomaterials are highly reactive due to their small size and large surface area. The extensive use of nanomaterials could lead to overproduction of reactive oxygen species, which cause deleterious effects on the plant's growth [75]. The variability and reactivity of nanomaterials are also a major concern. Besides this, it is very important to explore the suitability, sustainability, and feasibility of these new smart nano-fertilizers. There are some considerable concerns related to the toxicity, transport, dose-dependent response, and unintended environmental impacts on the biological systems limiting their acceptability in sustainable agriculture and medical sectors.

It has also been reported that the extensive use of nano-fertilizers has serious concerns in relation to food safety, food security, plants, and animals [76]. In addition, it is crucial to examine the level of toxicity of NPs in crops and understand the uptake, translocation, and utilization of nano-fertilizers and accumulation of nanomaterials in different plant tissues. The use of nanomaterials as fertilizers is continuously being explored for plant fortification and protection owing to their physico-chemical properties. In recent funded research projects and future research calls, there has been a greater focus on formulating and designing safer nanomaterials that can deliver effective, target-specific responses that are also environmentally friendly. Research in agriculture involving nanotechnology is still in its rudimentary stages but is advancing rapidly. However, there is an urgent need for a better understanding of mechanistic action, optimum dose, and various routes of application before its implementation at the field scale.

3.6 CONCLUSION

The increased and continuous use of conventional or chemical-based fertilizers, pesticides, and insecticides by farmers has resulted in decreased soil fertility, growth, physiology, biochemistry, and antioxidant efficacy of plants which ultimately results in less production of crops. The overuse of chemical-based fertilizers and their associated potential risks results in overexploitation of the agriculture ecosystem. In this scenario, the encouragement of the use of eco-friendly, sustainable, bio-compatible, and bio-inspired nano-fertilizers in plant growth promotion and in making the agro-ecosystem stable led to the development and ideal agriculture system. The formulation of bio-inspired nano-fertilizers has proven to change the current scenario of the agriculture sector by increasing the nutrient absorption efficiency, plant agronomic parameters, and productivity and resistance against various pathogens. Compared to synthetic fertilizers, bio-inspired nano-fertilizers provide more benefits to the environment and farmers. They may be described as a new approach to sustainable agriculture practices using NPs as inputs. Even though we have the right information about nanomaterials, assessment of toxicity levels is still necessary to avoid risks.

REFERENCES

1. Sajid M. Nanomaterials: types, properties, synthesis, emerging materials, and toxicity concerns. Current Opinion in Environmental Science & Health, 2021; 202; 100319.
2. Ikram M et al. Biomedical potential of plant-based selenium nanoparticles: a comprehensive review on therapeutic and mechanistic aspects. International Journal of Nanomedicine, 2021; 16: 249.
3. Zohra E et al. Potential applications of biogenic selenium nanoparticles in alleviating biotic and abiotic stresses in plants: a comprehensive insight on the mechanistic approach and future perspectives. Green Processing and Synthesis, 2021; 10(1): 456–475.
4. Zulfiqar F et al. Nanofertilizer use for sustainable agriculture: advantages and limitations. Plant Science, 2019; 289: 110270.
5. Okey-Onyesolu CF et al. Nanomaterials as nanofertilizers and nanopesticides: an overview. Chemistry Select, 2021; 6(33): 8645–8663.
6. Adisa IO et al. Recent advances in nano-enabled fertilizers and pesticides: a critical review of mechanisms of action. Environmental Science: Nano, 2019; 6(7): 2002–2030.
7. Goredema-Matongera N et al. Multinutrient biofortification of maize (*Zea mays* L.) in Africa: current status, opportunities and limitations. Nutrients, 2021; 13(3): 1039.
8. Aqeel U et al. A comprehensive review of impacts of diverse nanoparticles on growth, development and physiological adjustments in plants under changing environment. Chemosphere, 2021; 132672.
9. Ikram M et al. Foliar applications of bio-fabricated selenium nanoparticles to improve the growth of wheat plants under drought stress. Green Processing and Synthesis, 2020; 9(1): 706–714.
10. Iqbal M et al. Effect of silver nanoparticles on growth of wheat under heat stress. Iranian Journal of Science and Technology. Transactions A: Science, 2019; 43(2): 387–395.
11. Satti SH et al. Titanium dioxide nanoparticles elicited agro-morphological and physicochemical modifications in wheat plants to control *Bipolaris sorokiniana*. Plos One, 2021; 16(2): e0246880.
12. Akhtar N et al. Synergistic effects of plant growth promoting rhizobacteria and silicon dioxide nano-particles for amelioration of drought stress in wheat. Plant Physiology and Biochemistry, 2021; 166: 160–176.
13. Mustafa H et al. Biosynthesis and characterization of titanium dioxide nanoparticles and its effects along with calcium phosphate on physicochemical attributes of wheat under drought stress. Ecotoxicology and Environmental Safety, 2021; 223: 112519.
14. Mącik M, Gryta A, Frąc M. Biofertilizers in agriculture: an overview on concepts, strategies and effects on soil microorganisms. Advances in Agronomy, 2020; 162: 31–87.
15. Sharma N, Singhvi R. Effects of chemical fertilizers and pesticides on human health and environment: a review. International Journal of Agriculture, Environment and Biotechnology, 2017; 10(6): 675–680.

16. Yadav GK et al. Nano fertilizer in agriculture: a new trend of cost-effective farming. *In*: Research Trends in Agriculture Sciences. 2020. Naresh RK (ed.). Akinik Publications. New Delhi. pp. 3–10.
17. Butt BZ, Naseer I. Nanofertilizers. *In*: Nanoagronomy. 2020; Springer, Cham. pp. 125–152.
18. Gosavi VC et al. Synthesis of green nanobiofertilizer using silver nanoparticles of *Allium cepa* extract. IJCS, 2020; 8(4): 1690–1694.
19. Singh MD. Nano-fertilizers is a new way to increase nutrients use efficiency in crop production. International Journal of Agriculture Sciences, 2017; 0975–3710.
20. Qureshi A, Singh D, Dwivedi S. Nano-fertilizers: a novel way for enhancing nutrient use efficiency and crop productivity. International Journal of Current Microbiology and Applied Sciences, 2018; 7(2): 3325–3335.
21. Javed B et al. Biogenesis of silver nanoparticles to treat cancer, diabetes, and microbial infections: a mechanistic overview. Applied Microbiology and Biotechnology, 2021; 105(6): 2261–2275.
22. Adhikary R, Pal A. Geotextile and its importance in agriculture: a review study. International Journal of Bioresource Science, 2019; 6(2): 61–63.
23. Prasad R, Bhattacharyya A, Nguyen QD. Nanotechnology in sustainable agriculture: recent developments, challenges, and perspectives. Frontiers in Microbiology, 2017; 8: 1014.
24. Parry ML, Carter T. An assessment of the effects of climatic change on agriculture. Climatic Change, 1989; 15(1): 95–116.
25. Rajmohan KS, Chandrasekaran R, Varjani S. A review on occurrence of pesticides in environment and current technologies for their remediation and management. Indian Journal of Microbiology, 2020; 60(2): 125–138.
26. Mishra S et al. Heavy metal contamination: an alarming threat to environment and human health. *In*: Environmental Biotechnology: For Sustainable Future. 2019; Springer, Singapore. pp. 103–125.
27. Sharma VP et al. Advance applications of nanomaterials: a review. Materials Today: Proceedings, 2018; 5(2): 6376–6380.
28. He X, Deng H, Hwang HM. The current application of nanotechnology in food and agriculture. Journal of Food and Drug Analysis, 2019; 27(1): 1–21.
29. Younas A et al. Role of nanotechnology for enhanced rice production. *In*: Nutrient Dynamics for Sustainable Crop Production. 2020; Springer, Singapore. pp. 315–350.
30. Yaswanth PV et al. Scope and importance of nanotechnology in agriculture: a review. Plant Archives, 2021; 21(2): 29–34.
31. Chhipa H. Applications of nanotechnology in agriculture. *In*: Methods in Microbiology. 2019; Elsevier, London. pp. 115–142.
32. Duhan JS et al. Nanotechnology: the new perspective in precision agriculture. Biotechnology Reports, 2017; 15: 11–23.
33. Rai M, Ingle A. Role of nanotechnology in agriculture with special reference to management of insect pests. Applied Microbiology and Biotechnology, 2012; 94(2): 287–293.
34. Kumar S et al. Nano-based smart pesticide formulations: emerging opportunities for agriculture. Journal of Controlled Release, 2019; 294: 131–153.
35. Mikula K et al. Controlled release micronutrient fertilizers for precision agriculture – a review. Science of the Total Environment, 2020; 712: 136365.
36. Amin MA et al. The potency of fungal-fabricated selenium nanoparticles to improve the growth performance of *Helianthus annuus* L. and control of cutworm *Agrotis ipsilon*. Catalysts, 2021; 11(12): 1551.

37. Ikram M et al. Phytogenic selenium nanoparticles elicited the physiological, bio-chemical, and antioxidant defense system amelioration of huanglongbing-infected 'Kinnow' mandarin plants. Nanomaterials, 2022; 12(3): 356.
38. Kaur N et al. Selenium in agriculture: a nutrient or contaminant for crops? Archives of Agronomy and Soil Science, 2014; 60(12): 1593–1624.
39. Adnan M. Application of selenium a useful way to mitigate drought stress: a review. Biological Sciences Research, 2020; 3(1): 39.
40. Alves LR et al. Selenium improves photosynthesis and induces ultrastructural changes but does not alleviate cadmium-stress damages in tomato plants. Protoplasma, 2020; 257(2): 597–605.
41. Garousi F. Toxicity of selenium, application of selenium in fertilizers, selenium treatment of seeds, and selenium in edible parts of plants. Acta Universitatis Sapientiae Alimentaria, 2017; 10: 61–74.
42. Siddiqui SA et al. Effect of selenium nanoparticles on germination of *Hordeum vulgare* barley seeds. Coatings, 2021. 11(7): 862.
43. Gudkov SV et al. Production and use of selenium nanoparticles as fertilizers. ACS Omega, 2020; 5(28): 17767–17774.
44. Zahedi SM et al. Alleviation of the effect of salinity on growth and yield of straw-berry by foliar spray of selenium-nanoparticles. Environmental Pollution, 2019; 253: 246–258.
45. El-Saadony MT et al. The use of biological selenium nanoparticles to suppress *Triticum aestivum* L. crown and root rot diseases induced by *Fusarium* species and improve yield under drought and heat stress. Saudi Journal of Biological Sciences, 2021; 28(8): 4461–4471.
46. Djanaguiraman M et al. High-temperature stress alleviation by selenium nanoparticle treatment in grain sorghum. ACS Omega, 2018; 3(3): 2479–2491.
47. Rodríguez-Serrano C et al. Biosynthesis of silver nanoparticles by *Fusarium scirpi* and its potential as antimicrobial agent against uropathogenic *Escherichia coli* biofilms. Plos One, 2020; 15(3): e0230275.
48. Lal HM, Uthaman A, Thomas S. Silver nanoparticle as an effective antiviral agent. *In*: Polymer Nanocomposites Based on Silver Nanoparticles. 2021; Springer, Cham. pp. 247–265.
49. Sharma P et al. Silver nanoparticle-mediated enhancement in growth and antioxi-dant status of *Brassica juncea*. Applied Biochemistry and Biotechnology, 2012; 167(8): 2225–2233.
50. Kale SK, Parishwad GV, Patil ASHAS. Emerging agriculture applications of silver nanoparticles. ES Food and Agroforestry, 2021; 3: 17–22.
51. Hussain M et al. Green synthesis and characterization of silver nanoparticles and their effects on disease incidence against canker and biochemical profile in *Citrus reticulata* L. Nanoscience and Nanotechnology Letters, 2018; 10(10): 1348–1355.
52. Sultana T, Javed B, Raja NI. Silver nanoparticles elicited physiological, biochem-ical, and antioxidant modifications in rice plants to control *Aspergillus flavus*. Green Processing and Synthesis, 2021; 10(1): 314–324.
53. Ejaz M et al. Effect of silver nanoparticles and silver nitrate on growth of rice under biotic stress. IET Nanobiotechnology, 2018; 12(7): 927–932.
54. Durán N et al. Silver nanoparticles: a new view on mechanistic aspects on anti-microbial activity. Nanomedicine: Nanotechnology, Biology and Medicine, 2016; 12(3): 789–799.
55. Amooaghaie R, Saeri MR, Azizi M. Synthesis, characterization and biocompatibility of silver nanoparticles synthesized from *Nigella sativa* leaf extract in comparison

with chemical silver nanoparticles. Ecotoxicology and Environmental Safety, 2015; 120: 400–408.

56. Gohari G et al. Titanium dioxide nanoparticles (TiO$_2$ NPs) promote growth and ameliorate salinity stress effects on essential oil profile and biochemical attributes of *Dracocephalum moldavica*. Scientific Reports, 2020; 10(1): 1–14.

57. Mustafa N et al. Foliar applications of plant-based titanium dioxide nanoparticles to improve agronomic and physiological attributes of wheat (*Triticum aestivum* L.) plants under salinity stress. Green Processing and Synthesis, 2021; 10(1): 246–257.

58. Santhoshkumar T et al. Green synthesis of titanium dioxide nanoparticles using *Psidium guajava* extract and its antibacterial and antioxidant properties. Asian Pacific Journal of Tropical Medicine, 2014; 7(12): 968–976.

59. Ze Y et al. The regulation of TiO$_2$ nanoparticles on the expression of light-harvesting complex II and photosynthesis of chloroplasts of *Arabidopsis thaliana*. Biological Trace Element Research, 2011; 143(2): 1131–1141.

60. Rout GR, Sahoo S. Role of iron in plant growth and metabolism. Reviews in Agricultural Science, 2015; 3 : 1–24.

61. Marsh Jr H, Evans H, Matrone G. Investigations of the role of iron in chlorophyll metabolism. II. Effect of iron deficiency on chlorophyll synthesis. Plant Physiology, 1963; 38(6): 638.

62. Shebl A et al. *T*emplate-free microwave-assisted hydrothermal synthesis of manganese zinc ferrite as a nanofertilizer for squash plant (*Cucurbita pepo* L). Heliyon, 2020; 6(3): e03596.

63. Sreelakshmi B et al. Drought stress amelioration in plants using green synthesised iron oxide nanoparticles. Materials Today: Proceedings, 2021; 41: 723–727.

64. Mozafari A-a, Ghaderi N. Grape response to salinity stress and role of iron nanoparticle and potassium silicate to mitigate salt induced damage under in vitro conditions. Physiology and Molecular Biology of Plants, 2018; 24(1): 25–35.

65. Askary M et al. Effects of iron nanoparticles on *Mentha piperita* L. under salinity stress. Biologija, 2017; 63(1): 65–75.

66. Yruela I. Copper in plants. Brazilian Journal of Plant Physiology, 2005; 17: 145–156.

67. Abbasifar A, Shahrabadi F, Valizadeh Kaji B. Effects of green synthesized zinc and copper nano-fertilizers on the morphological and biochemical attributes of basil plant. Journal of Plant Nutrition, 2020; 43(8): 1104–1118.

68. Piper C. Investigations on copper deficiency in plants. The Journal of Agricultural Science, 1942; 32(2): 143–178.

69. Ngo QB et al. Effects of nanocrystalline powders (Fe, Co and Cu) on the germination, growth, crop yield and product quality of soybean (Vietnamese species DT-51). Advances in Natural Sciences: Nanoscience and Nanotechnology, 2014; 5(1): 015016.

70. Hafeez A et al. Potential of copper nanoparticles to increase growth and yield of wheat. Journal of Nanoscience with Advanced Technology, 2015; 1(1): 6–11.

71. Zhao L et al. Nano-biotechnology in agriculture: use of nanomaterials to promote plant growth and stress tolerance. Journal of Agricultural and Food Chemistry, 2020; 68(7): 1935–1947.

72. Chikte R et al. Nanomaterials for the control of bacterial blight disease in pomegranate: quo vadis? Applied Microbiology and Biotechnology, 2019; 103(11): 4605–4621.

73. Lopez-Lima D et al. The bifunctional role of copper nanoparticles in tomato: effective treatment for *Fusarium* wilt and plant growth promoter. Scientia Horticulturae, 2021; 277: 109810.

74. Iqbal M et al. Assessment of AgNPs exposure on physiological and biochemical changes and antioxidative defence system in wheat (*Triticum aestivum* L) under heat stress. IET Nanobiotechnology, 2019; 13(2): 230–236.

75. Dev A, Srivastava AK, Karmakar S. Nanomaterial toxicity for plants. Environmental Chemistry Letters, 2018; 16(1): 85–100.

76. Iavicoli I et al. Nanotechnology in agriculture: opportunities, toxicological implications, and occupational risks. Toxicology and Applied Pharmacology, 2017; 329: 96–111.

4 Natural Biopolymer Nanomaterials in Biotic Stress Management

Seyedeh-Somayyeh Shafiei-Masouleh,
Hamed Hassanzadeh Khankahdani, and
Santosh Kumar

CONTENTS

4.1 INTRODUCTION

Feeding the ever-increasing world population, having sustainable agriculture, use of nanotechnology in agriculture, looking for natural compounds to manage agriculture, reducing environmental pollution, and human health are together of particular concern for today's world. Although nanotechnology has revolutionized various industries such as medical, engineering, and electronics, agriculture needs to move forward more rapidly to solve these specific problems. Based on research and experience, a combination of biogenic substances and nanotechnology could help defeat these issues [1]. Global warming has caused continuous change in the behavior of microorganisms, and there are numerous problems in controlling them. On the other hand, antimicrobial compounds lose their activity rapidly as well as polluting the environment. Therefore, the use of antimicrobials (fungicides, bactericides, and so on) with natural structures may remove these problems, especially when integrated into nano-science and technology [2]. Particles with sizes less than 100 nm in even one dimension are named nanomaterials [3]. The special importance of nanotechnology is

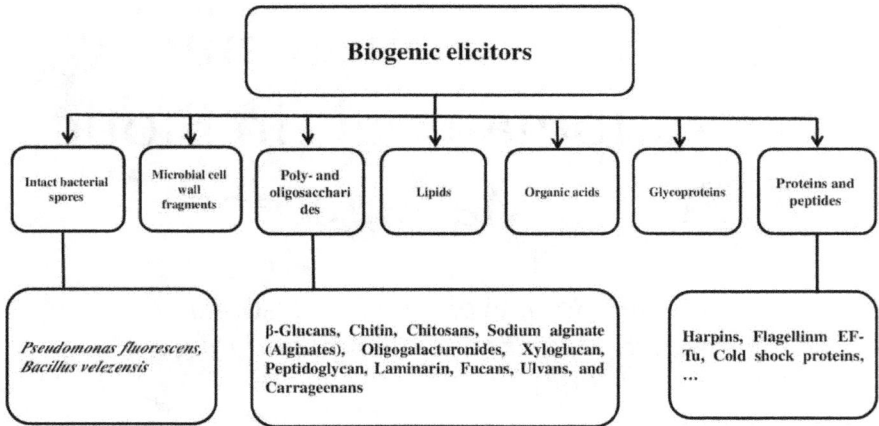

FIGURE 4.1 Some biogenic elicitors trigger plant signal transduction pathways.

Source: adopted from [5].

in achieving sustainable agriculture through an increase in crop protection as well as in crop performance. In the technology, there has been interest in certain properties for sustainability of biopolymers such as alginate, chitosan, and glucan . Properties of biopolymers that may be noticed are biocompatibility, biodegradability, low toxicity, and desirable functionality. They have dual roles in agriculture, including plant growth and development, and plant protection against biotic and abiotic agents. Finally, at nano-size levels, they have more suitable properties [4].

Signal transduction pathways in plants are needed to activate plant immunity against biotic and abiotic stresses to induce plant resistance, and all of them are triggered by elicitors. The elicitors are different types of exogenous and endogenous compounds which can be categorized as shown in Figure 4.1. Elicitors with nature of proteins and peptides that have microbial origin can be effective, in comparison to organic compounds, which have simple structures.. However, problems such as exposure to ultraviolet (UV) radiation on the plant surface or various plant proteases, after penetrating plant tissues and cells, limit their use, and one method to protect them is to apply biopolymers. That said, this complexation may create synergetic effects of elicitors, because some of the biopolymers, such as chitosan and sodium alginate, play roles as elicitors [5].

Today, sustainable agriculture is of worldwide interest, and natural compounds can be considered to be one of the key components. Nano-size biopolymers with their unique properties have been reported for their beneficial effects on different plants in terms of growth and development, photosynthesis, yield, and plant protection against biotic and abiotic stresses. This chapter reviews various roles of natural biopolymers, and particularly their nanomaterials in regard to the management of biotic stress in agriculture. It introduces available nanomaterials and their biopolymer structures and effective elements in plant responses, and plant–pathogen/pest reactions at the level of molecules and organs/tissues/cells.

4.2 BIOTIC STRESS

To accumulate assimilation and have an appropriate reproductive stage and maturity, plants must cope with a wide range of biotic and abiotic stresses. However, when the issues are complicated, plant assimilates or nutrients are the primary targets for pathogens. Nevertheless, plants may be able to be disease-free. Plants prevent pathogens through many mechanisms, such as waxy cuticles and constitutively produced toxic compounds [6]. Greenhouse gases and undesirable environmental conditions have affected CO_2 concentrations, temperature, and water availability, and all of them affect plant responses to stressful conditions, including pathogens. When the environmental condition is favorable, even susceptible plants as the hosts may not be affected by a virulent pathogen [7]. Some plant resistance pathways include pattern-triggered immunity, effector-triggered immunity, defense hormone networks, and RNA interference, which are all affected by environmental conditions as well as environmental impacts on pathogens, including the production of toxins and virulence proteins, pathogen reproductive features and its survival, which are affected by temperature and humidity [7].

How plants are affected by biotic agents, especially fungi and bacteria, depends on how these agents enter and invade plants. For instance, pathogens like *Verticillium dahlia* are vascular pathogens and some, like *Botrytis cinerea*, are non-vascular [8]. Most Gram-negative bacteria in the Pseudomonadaceae and Enterobacteriaceae colonize the apoplast and cause virtual symptoms such as rot, spots, wilts, cankers, and blights in crops. These bacteria are small and enter the plant apoplast through stomata and other natural pores [9].

In this chapter, we will notice some generalities of plant–pathogen interactions. This helps us to understand the effects of biopolymers that are naturally biosynthesized in plants against pathogens or can be used exogenously for sustainable agriculture as well as the healthy environment from the viewpoint of agriculture and horticulture.

4.2.1 PLANT INTERACTIONS

Plant pathogens create problems for crop yield and food quality and security. On the other hand, plants with various natural compounds in cell walls and enzymes protect themselves against pathogens; this is called induced resistance. Plants use more oxygen against pathogens and this produces more reactive oxygen species (ROS); however, this is toxic for plants as well as pathogen cells, as it causes photo-oxidative damage to bio-molecules and internal cellular structures. Furthermore, plants with non-enzymatic compounds, such as phenolic compounds, flavonoids, lignins, accumulation of tannins and phytoalexins, and enzymatic responses to phenylalanine ammonia-lyase (PAL), polyphenol oxidase and antioxidant enzymes like superoxide dismutase, catalase, peroxidases, and glutathione reductase, attack pathogen cells. That said, plants with antioxidant compounds protect themselves against toxins produced against pathogens, and secondary metabolites like phytoalexins in infection sites make the hypersensitive response. Moreover, polyamines are important compounds in plants and play a role in cellular metabolism and alter pathogen activities [10].

Some pathogens are insect-borne; when plants are infected they experience two simultaneous injuries. Therefore, plants must deploy broad-spectrum mechanisms against both. Plants have multi-stress resistance-related genes, many of which play roles in innate immunity and phyto-hormone signaling. The two best phytohormone signalings are related to jasmonate and salicylic acid. Therefore, understanding genome editing and chemical modulators may help to find sustainable ways of coping with insect-borne diseases [11]. At various levels, such as organs/tissues or cells and molecular levels, plants counteract pathogens. Some of these plant actions are specific to particular plants, and some apply generally to a wide range of plants.

4.2.1.1   Plant Tissues and Cells

Plants have specific differences from animals; their bodies and mechanisms enable them to withstand unfavorable conditions. For example, they have waxy compounds on their leaves, stems, and fruit to cope with entering pathogens and insects or tolerate undesirable conditions of abiotic stress. As known, plasmodesmata are plant cell membrane pores (intercellular channels) that have a role in connections between adjacent cells to mediate symplastic communication. These pores play a part in regulating the molecules essential for plant development and stress responses; however, some pathogens may use these pores to invade plant cells. Plants also have various corresponding defense mechanisms to regulate these pores (through the appropriate modulation of callose deposition and plasmodesmal permeability) to impede pathogen spread [12].

4.2.1.2   Molecular Levels

The interaction between plants and pathogens in their life cycles has created the co-evolutionary cycles known as "Red Queen" dynamics and changed both genomes [13]. An immune system for plants requires a high energy cost to make plants powerful against microbial pathogens. Plant defense mechanisms include processes such as DNA methylation, transformation of histone density and variants, and many epigenomic changes related to plant immunity [14]. Both endophytes and fungal pathogens that evade plants convert chitin to chitosan. Furthermore, they protect chitin as well as fungal cell walls from plant chitinases with their LysM effectors. Chitinases, fungal chitin deacetylases, chitosanases, and effectors together play a role in the fungal colonization of plants, and in whether a plant is immune or not [15]. After invading plant tissues pathogenic bacteria proliferate in extracellular spaces, and one plant defense system is the accumulating lignin (a type of polymer) that strengthens cell walls to protect plants against pathogen spread [16].

Plants are unlike mammals and have no mobile defender cells (like white blood cells) and a somatic adaptive immune mechanism [6, 17]. Therefore, they need to rely on the innate immunity of cells and systemic signals of infection sites. These responses are of two types: pattern-triggered immunity and effector-triggered immunity, the first and the second modes of immunity, respectively. The first is started by recognizing molecular patterns that are common to many types of microbes, and the second is based on recognizing pathogen effectors [18]. The second system is prolonged because it is activated in the cytosol while the first is activated on the cell membrane. Pattern recognition receptors are the proteins that recognize two types of patterns, pathogen-associated molecular patterns and damage-associated molecular patterns. Both of these are elicitors. Therefore, elicitors as biological strategies have an important role in the plant defense system and induce plants to produce metabolites and create plant resistance [17–19]. This means that plants have an effective immune system, and each cell individually recognizes invaders. In this type of system, there is an interaction between a putative plant-derived receptor and a corresponding molecule of a pathogen, which is called an elicitor. The productions of pathogens' elicitors are affected by "avirulence" (*Avr*) genes vis-à-vis *R* genes as plant resistance genes, which are receptor molecules for recognizing specific elicitors.

Finally, gene-for-gene resistance is initiated by the interaction of *R* and *Avr*. When a pathogen enters host tissue, it secretes specialized molecules, including virulence factors, and uses plant resources, and finally, disease occurs or does not [6, 17, 19]. From a specific *R–Avr* interaction, very large intracellular and intercellular changes occur, and the phosphorylation state and large ion fluxes, especially calcium ions occur in cells in and around the recognition site; this is early signal transduction. Then salicylic acid as a signaling molecule is induced to make subsequent systemic plant defenses. Producing ROS and nitric oxide (NO) promotes a hypersensitive response synergistically; this is called localized programmed cell death. The production of reinforced walls (with callose and lignins) restricts pathogen invention. Furthermore, some ROS are toxic for pathogens inside antimicrobial compounds like phytoalexins and digestive enzymes like chitinases and glucanases. All of the processes mentioned are part of the plant defense mechanism [6, 17]. The hypersensitive response in plants against pathogen invasion includes both electrolyte leakage and rapid and highly localized programmed cell death [20].

The most important defense reactions of plants against invading pathogens, including fungi, bacteria, nematodes, and insects, are phytohormones (especially salicylic acid, jasmonic acid, and ethylene) and secondary metabolites (such as phenylpropanoids and terpenoids). For instance, phenylpropanoids produce phenyl-alanine, an aromatic amino acid, which is an intermediate in the synthesis of salicylic acid. Salicylic acid plays a role in programmed cell death and then protects plants against pathogen spread [21].

Another plant mechanism against pathogens is RNA transmission, and small RNAs regulate trans-kingdom RNA silencing in plant immunity as well as the well-known RNAi mechanisms in plants and fungi [8]. Glutathione S-transferases (GSTs) are plant enzymes that are encoded by a large gene family. A specific feature of the enzyme is its induction and up-regulation by a wide range of stressful conditions, including pathogenic stresses. Some important roles of the enzyme are detoxifica-tion of toxic compounds through their conjugation with glutathione, reducing the oxidative stress effects, and a role in hormone transport. It is documented that some GSTs with glutathione peroxidase activity detoxify toxic lipid hydroperoxides; some have a role in the intracellular transport of auxins, and/or are the receptor proteins of salicylic acid, and induce a systemic resistance response. That said, plant treatment with beneficial microbes shows these processes [22]. Phytooxylipins are other plant defense systems that are induced against stressful conditions and make stress-related signaling pathways as well as having non-signaling roles. They include ROS and electrophile species and based on their chemical structures induce gene expression related to the defense system [23].

Another plant defense reaction is resistosomes (higher-order oligomeric complexes), which are a complex between nucleotide-binding and leucine-rich repeat (NLR) proteins (intracellular immune receptors) upon direct or indirect recognition of effector proteins [24]. Potato and soybean are two important world crops, and both are usually subjected to *Phytophthora* pathogens, a persistent threat. Currently, disease resistance breeding and chemical pesticides are the best strategies to cope with the disease. It has been explored that *Phytophthora*-derived phosphatidylinositol 3-phosphate (PI3P), as a novel control target, uses proteins (secreted antimicrobial peptides and proteins) to bind this lipid to the surface of the *Phytophthora* pathogens; therefore, transgenic plants with PI3P are suitable for sustainable production [25].

It is said that even plants with volatile organic compounds respond to attackers. For instance, tea plants with a terpenoid volatile (*E*)-nerolidol elicit plant responses against the piercing herbivore *Empoasca* (*Matsumurasca*) *onukii* Matsuda and the pathogen *Colletotrichum fructicola*. This terpenoid triggers the activation of a mitogen-activated protein kinase, induction of jasmonic acid and abscisic acid signaling, H_2O_2 burst, and finally defense responses against insect pests and fungi [26].

4.3   NATURAL BIOPOLYMERS

For sustainable agriculture and to maintain a healthy environment and human life, biopolymers may be regarded as beneficial. Biopolymers are natural sources; they are polymers synthesized through chemical methods or by living organisms. However, these polymers are mechanically poor; they are subsequently strengthened with fillers and nanofillers to create a broad range of applications. Therefore, by making the composites named green composites, individuals could reach their target of sustainable agriculture. Polylactic acid, chitosan, alginate, and natural rubber are some of these polymers, and in particular chitosan and alginate have a special importance in agriculture. It is reported that adding polyphenols such as tannic acid, gallic acid, and ferulic acid to polysaccharides like chitosan may increase its antimicrobial activity [27].

4.3.1   NATURES AND STRUCTURES

Based on the inspiration of plant nature, exogenous elicitors may replace pesticides and other microbicides, especially when they are only needed to be used in micrograms with long-term efficiency, and are beneficial for human and environmental health. One elicitors that is frequently reported is carbohydrates, especially poly- and oligosaccharides. Polysaccharides such as chitin, chitosan, β-glucans, oligogalacturonides, alginates, xyloglucan, peptidoglycan, laminarin, fucans, ulvans, and carrageenans are common carbohydrate elicitors; their activities as elicitors vary and depend on certain factors (Table 4.1). These elicitors act following both patterns mentioned above. Carbohydrate elicitors have different origins, such as plant-derived,

TABLE 4.1
Some Important Carbohydrates with Elicitor Roles in Plants Against Pathogens

Carbohydrates	Effective factors	Defense responses	Pathogen agents
β-glucan	Degree of polymerization, glycosidic bond type, branching characteristics, and chemical modification	Calcium influx, ROS burst, medium alkalinization, mitogen-activated protein kinase (MAPK) activation, SA accumulation, defense-related gene expression, defense enzyme activation, and phytoalexin production	*Botrytis cinerea, Plasmopara viticola,* and tobacco mosaic virus (TMV)
Chitin and chitosan	Molecular mass, degree of acetylation and chemical derivative attached	Calcium influx, plasma membrane H⁺-ATPase inhibition, ROS burst, MAPK activation, synthesis of phytohormones, defense enzyme activation, pathogenesis-related protein accumulation, production of phytoalexin, and callose deposition	Fungi more than bacteria, and Gram-positive bacteria more than Gram-negative bacteria
Oligogalacturonides	Degree of polymerization and chemical derivative	ROS burst, callose deposition, MAPK activation, defense enzyme activation, and defense-related gene expression	–
Carrageenans	Degree of substitution, degree of sulfation	–	–
Ulvans	Degree of polymerization and sulfate substituent	–	

Note: ROS, reactive oxygen species; SA, salicylic acid.

algae-derived, and fungal-derived. Their effects depend on plant species and stage of plant growth and development, elicitor types, concentrations used, and conditions [18].

Chitin biopolymers are the second most common polysaccharides after cellulose, which are abundant on the earth. This polymer and its derivative mean chitosan could induce a plant defense system, and their effects depend on the degree of acetylation. Chitin is found in fungal cell walls, shrimp, and crabs. Plants perceive a fungal invasion based on processes, including plant–fungal interactions, roles of chitin hydrolases in fungi, and chitinase production in plants [28]. Chitin is an important polymer in fungal cell walls and is an effective elicitor in the plant immunity system [29].

One biopolymer that may transform agriculture is chitosan, which is a biocompatible, biodegradable, and non-toxic compound. It can play a part in the controlled release of agrochemicals and in protection of plants against pests and pathogens, soil fertilization, delivery of genetic materials, increasing tolerance to abiotic stresses, and finally plant growth [3]. Chitosan, a hetero-amino-polysaccharide, is a biopolymer that has more applications than other biopolymers such as chitin, starch, gelatin, cellulose, and glucans. This is because it has a unique chemical and physical structure, including the availability of amino functional groups ($-NH_2$), high responsiveness to pH changes, and capacity for size modulation [1]. It is said that the first elicitor activity of chitosan was identified based on defense enzymes such as chitinase and fruits such as strawberries. Chitosan was reported to be effective in the expression of a thousand or more genes. Its activity also depends on the concentrations used as well as other physical and biochemical properties, mentioned above. It has been documented that chitosan, with a higher degree of deacetylation, because it has more positively charged amino groups, could have more electrostatic interactions with different substrates, and thus may have greater antimicrobial activity at various ranges of pH [30]. In Figure 4.2, the mechanism of antimicrobial activities of chitosan is categorized in direct and indirect ways [31].

Alginate is a widely used biopolymer extracted from seaweed and used for coating (encapsulation) bio-control bacteria. This polymer is biocompatible and biodegradable, and supports bacteria function in the long term. In modern agriculture, although this polymer is protective for beneficial bacteria, it can cause targeted release of bio-bacteria and have a role in the production of organic crops [32]. Furthermore, alginate oligosaccharide, a degradation product of alginates, is reported to have a role in pathogen-associated molecular patterns, and then to activate, as elicitor, the plant defense system against the pathogen. In food packaging, it inhibits the expression of ABA signaling pathway genes and enables fruits to have a long shelf life. The degree of polymerization is an important factor in alginate oligosaccharide activity as an elicitor: high and medium degrees are more effective [33].

4.3.2 ACTIVITIES AND REACTIONS

The use of biological control is one of the favored trends in modern agriculture. Bio-control agents such as *Pseudomonas* and *Bacillus* as bio-fertilizers are used in agriculture and can enhance the desired characteristics of soils and plant availability to nutrients as well as improving the efficiency of organic and mineral fertilizers. However, their liquid and solid formulations have problems due to their short shelf

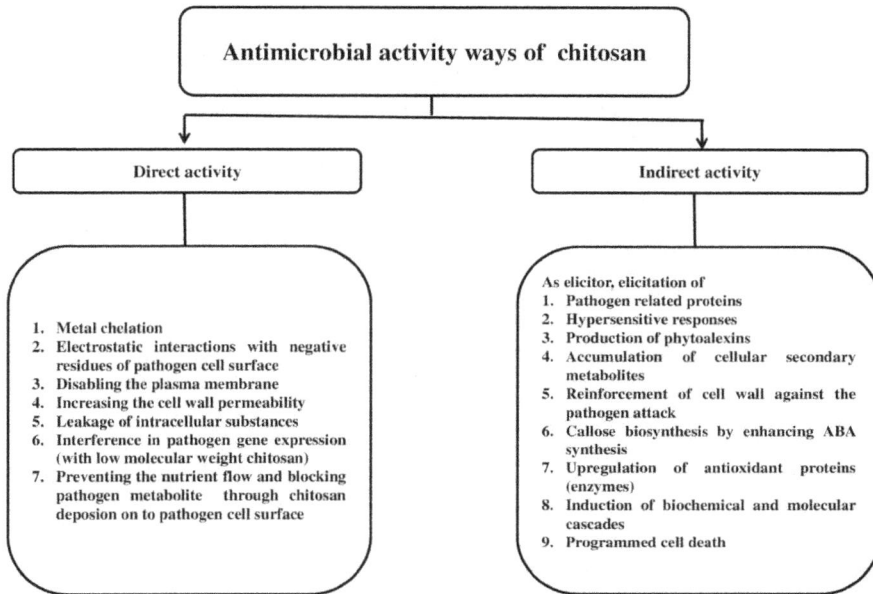

FIGURE 4.2 Two ways of antimicrobial activities of chitosan in plants.

Source: adapted from [31].

life, and encapsulation can remove these issues [34, 35]. Therefore, it is necessary to guarantee the survival of bio-control bacteria in host plants and this can be done through encapsulation of bio-controls by biopolymers such as alginate. Different methods are used to coat bio-control, such as extrusion, spray drying, and emulsion [34]. A novel agro-formulation of *Trichoderma viride* spores and calcium alginate microspheres has been introduced [36]. The researchers encapsulated the spores inside the microspheres. Both of them protect plants against biotic agents and affect plant nutrition. These two biological and chemical agents were examined. The authors observed more germination of the spores; encapsulation made a supportive environment for *T. viride* growth, and this microorganism could grow and penetrate out of microspheres. The researchers reported the rate of penetration out of microspheres depended on the calcium ion concentrations: greater concentration made the alginate network structure stronger.

In another study, *Pseudomonas chlororaphis* strain VUPF506, a bio-control against *Rhizoctonia* disease in potato plants, with alginate was encapsulated [37]; whey protein was used to strengthen the microcapsule structure. The researchers reported that this strengthened microcapsule was more efficienct and had a better effect on *P. chlororaphis* to control pathogens. In addition, micro-encapsulation techniques for *Pseudomonas fluorescens* VUPF506, as probiotic bacteria in soil, including alginate combined with whey protein carboxymethyl cellulose, and peanut butter were studied [38]. The authors recommended the first encapsulating compound for the controlled releasing probiotic against *Rhizoctonia solani* potato plants *in vivo*, in terms of controlling the pathogen, and for amino acids of whey protein-promoting plant growth (as a fertilizer).

Humans threaten the environment and their health using chemical pesticides, so green crop protections is of interest. One such green protection is the use of microbial proteins, which can activate plant defense systems and induce resistance to plant pathogens. These proteins can act as elicitors for plants; however, their capacity against adverse environments is limited, and may not affect the targeted receptors of plants. Therefore, elicitors used in agriculture need to be protected. One such elicitor is heat-resistant FKBP-type peptidylprolylcis-*trans* isomerase (PPIase) from *Pseudomonas fluorescens*, which has eliciting activity against some plant pathogens [5]. PPIase with 70% sodium alginate and 20% bovine serum albumin was encapsulated [5] and micro-particles with 10% PPIase per total volume were produced. The eliciting activity of the encapsulated PPIase compared to free PPIase in three different "plant–pathogen" models ("tobacco–tobacco mosaic virus," "tobacco–*Alternaria longipes*," and "wheat–*Stagonospora nodorum*" model systems) was examined and doubled activity of the encapsulated one was found. A new protein elicitor from *Bacillus subtilis* BU412, with effects of hypersensitive response and systemic acquired resistance, was reported [39]. It was used on tobacco leaves and triggered early defense events such as the generation of ROS (H_2O_2 and O_2-) and the induction of defense enzymes, including polyphenol oxidase, peroxidase, superoxide dismutase, and PAL. It was named AMEP412. It was observed that it can stimulate plant systemic resistance against *Pseudomonas syringae* pv. *tomato* DC3000.

Marine algae polysaccharides have been observed with different characteristics such as bio-stimulants, elicitors (as a natural plant defense system), and bio-fertilizers in agriculture and horticulture [40]. Numerous pathogens, including fungi, bacteria, nematodes, and viruses, embroil plants and then to survive, plants react with defense systems. Hormones, including salicylic acid, jasmonic acid, and ethylene, play a role in the activation of the system and molecular signals. Recent studies have shown that natural compounds with an environmentally friendly nature such as alginates can induce plant defense systems as well as plant growth. Alginates is the term used for salts of alginic acids; they are categorized as polysaccharides, and it has been reported that the polysaccharides and their oligosaccharides cause an expression of a salicylic acid-dependent defense pathway [41]. It is reported that the alginates extracted from seaweeds such as *Himanthalia elongate* can have antibacterial effects against Gram-positive and Gram-negative bacteria as well as mechanical and thermal properties [42]; therefore, they can be used in antimicrobial food packaging. Alginate oligosaccharide-induced resistance to *Pseudomonas syringae* pv. *tomato* DC3000 in *Arabidopsis thaliana* (wild-type and salicylic acid (SA)-deficient mutant (*sid2*)) was evaluated [43]. It was found that alginate oligosaccharide induced resistance in both; however, the types of effect were different. In the wild type, it decreased disease index and bacteria colonies, boosted ROS and NO, the expression of resistance gene *PR1*, and the content of SA); however, in the mutant type, the decreased rate of the disease index was more than in the wild type; the gene expressions of *recA* and *avrPtoB* were two and four times lower than the wild type; and alginate oligosaccharide induced disease resistance after 3 days by exciting SA pathways.

Sodium alginate/*O*-carboxymethyl chitosan hydrogels (*O*-CMCh) for slow-release fertilizer of urea using calcium chloride as a cross-linker were prepared [44]. It has been reported that these hydrogels exhibited more antimicrobial activity against some pathogenic bacteria and fungi in soil compared to *O*-CMCh. In a study, alginate controlled Bayoud disease in date palms (*Fusarium oxysporum* f. sp. *albedinis* (Foa)) through over-expression of PAL in roots and genes of oxidative events (superoxide dismutase and lipoxygenase) [45]. Sodium alginates extracted from different species of brown seaweeds (*Fucus spiralis* and *Bifurcaria bifurcata*) may have shown elicitor activity with different intervals [46]; however, both produced more PAL activity and total polyphenol content in date palm seedling roots, as a symptom of inducing the defense system. Alginate that was isolated from *Bifurcaria bifurcata* (an alga and its oligoalginate derivatives) in tomato seedlings to investigate their elicitor capacities based on the levels of PAL and polyphenol content in the leaves was examined [40]. It was observed that both elicitors are similar for PAL and triggering plant defense responses; however, oligoalginates was more effective as regards levels of polyphenols. Generally, both elicitors could be effective in phenylpropanoid metabolism and cope with plant pathogens.

Up-regulations of plant defense enzymes/genes under the application of biopolymers that could protect plants against pathogens has been reported [4]. Alginate oligosaccharide for the treatment of harvested kiwifruit stored at 25°C against gray mold, blue mold, and black rot was used [47]. Alginate oligosaccharide, which is a biological carbohydrate, from sodium alginate, had various polymerizations. The treatment inhibited pectin solubilization and gene expression of *pectin methylesterase* and *polygalacturonase*. Furthermore, the levels of total phenols and flavonoids in fruits increased, and the treatment enhanced fruit quality and shelf life.

For optimum usage of biopolymers to protect plants against biotic agents, researchers examined various formulations and methods such as hydrogels, films, beads, and so on. For instance, it was stated that sodium alginate films mechanically are relatively poor and have no antibacterial properties, and cannot be used for packaging fresh-cut fruits [48]. The researchers recommended thymol/sodium alginate composite films. Thymol is a phenol compound with antibacterial effects. The authors showed that these films have highly scavenging activity against 1,1-diphenyl-2-picrylhydrazyl radicals, inhibitory effects against *Staphylococcus aureus* and *Escherichia coli*, and finally maintain fruit quality and weight. Gummosis gel beads with two lyophilized yeast strains, including *Pichia guilliermondii* (5A) and *Candida oleophila* (13L), and sodium alginate to control *Penicillium digitatum*, the causal agent of green mold disease on citrus fruit under postharvest conditions, was prepared [49]. These gel beads especially with both yeast species inhibited fruit decay and prolonged their storage shelf life under 4°C. Edible films for food packaging with sodium alginate and essential oils of four medicinal plants (*Rosmarinus officinalis*, *Artemesia herba alba* Asso, *Oncimum basilicum* L., and *Mentha pulegium* L.) were produced [50]. It was reported that essential oils were dispersed uniformly; the films had antibacterial effects against six pathogenic bacteria as well as the antioxidant capacity of the film. The authors reported that essential oils improve the thermal and barrier properties of alginate films.

In a study, EPS66A was investigated [51]. EPS66A is a polysaccharide extracted from an unidentified species of bacterium *Streptomyces*; it was applied to tobacco leaves against the tobacco mosaic virus. The authors observed the polysaccharide induced two defense systems, including hypersensitive response, increase in NO, SA, and oxidative burst and, at the second level, callose deposition to prevent virus invasion, and an increase in expression of pathogenesis-related genes.

Many roles of chitosan, a plant defense modulator, have been documented. For instance, chitosan can penetrate the high-fluidity plasma membrane and increase the biosynthesis of intracellular oxygen species (ROS). It also plays a role in the overexpression of the tryptophan-dependent auxin biosynthesis pathway, and the accumulation of auxin in roots [15]. It is documented that chitosan as a biotic elicitor leads to induced systemic resistance in plants. The chitosan function for defense gene activation is related to the accumulation of different enzymes and stress-specific production and accumulation of pathogenesis-related (PR-) proteins. Among the most important PR proteins are thaumatin-like proteins, β-1,3-glucanases (PR-2), non-specific lipid transfer proteins, chitinases (PR-8), ribosome-inactivating proteins (PR-10), peroxidase defenses, oxalate oxidase, thionins, and oxalate-oxidase-like proteins, which are reported for their inhibition against *Fusarium oxysporum*, *Pestalotia* sp., *Erwinia amylovora*, and *Pseudomonas solancearum* [52].

4.3.3 NATURAL BIOPOLYMER NANOMATERIALS

Nanotechnology has been considered a promising technology to cope with pests and diseases in agriculture and deals with particles at dimensions of 1–100 nm. Furthermore, colloidal particles (10–1000 nm) are also counted as nano-particles in agriculture. Special properties of nano-particles, such as high dispersion, wettability (to control pesticide runoff), and even properties that are not observed in bulk (for instance, silver has antimicrobial features only at nano-sized dimensions), are of particular interest [53]. Islam used nano-based deltamethrin, which is an insecticide, and promoted its solubility and dispersion [53]. Furthermore, they said that nano-formulation leads to more bio-efficacy against *Trialeurodes vaporariorum* (greenhouse whitefly) than bulk formulations. It was stated that some properties of nano-formulation of insecticides, such as photocatalytic property, optimized release capability, sustainable and constant release, targeted delivery, and prolonged persistence time of nanomaterials, are the reasons for the high efficiency of nano-based pesticides.

Based on the above-mentioned reasons, the combination of biopolymer usage and nanotechnology has been of interest to researchers. For instance, an alginate–gelatin nano-composite was used to make beads of two strains of *Pseudomonas fluorescens* (VUPF5 and T17–4) [2], which are beneficial bacteria for plant growth. These methods were important steps to conserve these beneficial bacteria against adverse conditions of the rhizosphere. In another study, sodium alginate from *Macrocystis pyrifera* (algae) was extracted [54]. Then the authors enriched sodium alginate–gelatin micro-capsules with carbon nano-tubes and SiO_2 nano-particles, and finally, they encapsulated *Bacillus velezensis* with the enriched micro-capsules. They used the micro-capsules against pistachio gummosis (*Phytophthora drechsleri*), and for growth parameters, and found desirable results. Furthermore, they reported that the survival rate of the encapsulated bacteria was high and up to 1 year of storage.

Nano-encapsulated *Bacillus subtilis* Vru1 was reported to control *Rhizoctonia solani* in beans [55]. The authors showed that all coating compounds (sodium alginate, starch, and bentonite) were more effective than free *Bacillus subtilis*, and when nano-encapsulation was combined with titanium dioxide nano-particles, the efficiency was even greater. The researchers observed efficiencies through the number of bacteria, promoting vegetative growth, and production of indole-3-acetic acid. Kanagaraj et al. [56] synthesized alginate-based eco-friendly, biodegradable nanocomposite films of silver nano-particles isolated from *Celosia cristata* leaf extracts (plant-derived biosynthesis). They used the synthesized alginate–silver (Alg-Ag) nano-composite films against some food-borne pathogens, i.e., *Staphylococcus aureus*, *Salmonella typhimurium*, and *Clostridium perfringens*. They observed that the films enhance food quality and storage stability as well as eliminate harmful microbes.

Chitosan has a special and interesting molecular structure that leads to its broad-spectrum activities. It is a positively charged linear molecule with a random arrangement of β-(1–4)-linked D-glucosamine and *N*-acetyl-D-glucosamine, and a derivative of chitin. Chitosan is produced from the deamination of chitin. Because its positive charge can link to negatively charged structures in pathogens (even in Gram-positive bacteria) and plants, then various biochemical and molecular changes in plant defense systems and pathogen structures are induced, including molecular, bio-chemical, and organ levels. It is reported that nano-formulations of chitosan are more efficient because of high surface area, higher solubility, and bio-stimulation activity compared to bulk forms [3, 31]. Defense responses of plants against pathogens include the production of ROS, biosynthesis of ABA, salicylic acid, jasmonic acid, membrane depolarization, hypersensitive responses, callose formation, expression of defense-related genes, and programmed cell death, and chitosan was reported to induce and enhance these responses. Furthermore, modification of chitosan as loading with metal ions such as Ag^+, Cu^{2+}, or Zn^{2+}, as well as creating nano-particles, promoted the anti-microbial activity of chitosan [3].

Based on evidence, nano-pesticides with greater efficiency and specificity target, reduce the use of pesticides, and then control pollution of the environment and non-targeted organisms. The composite structure of nano-pesticides with biopolymers like carboxymethyl chitosan has been reported to be promising for successful pesti-cide delivery and pest control [53]. It is accepted that nano-particles of chitosan for antimicrobial activities have greater efficiency compared to natural polymer chitosan [52]. It was investigated that chitosan and chitosan nano-particles had effects on *Fusarium andiyazi* (wilt disease in tomato) [52]. The researchers reported maximum inhibition for radial mycelial growth under the highest concentration of antifungal compounds, and more so with nano-particles – approximately 20% higher. They observed that the up-regulation of β-1,3-glucanase, chitinase, PR-1, and PR-10 genes under nano-particles was approximately 1.5–2 times more than chitosan. That said, the up-regulation of the transcript profile of superoxide dismutase under nano-particles treatment was 4.5 times more than chitosan. Both biopolymers were studied as beneficial bio-controls.

Sathiyabama and Charles [57] isolated cell wall polymer (chitosan) from *Fusarium oxysporum* f. sp. *lycopersici*, and, based on their cross-linking with sodium tripolyphosphate, synthesized nano-chitosan. They used the nano-particles on *F. oxysporum* f. sp. *lycopersici*-infected tomatoes and observed that wilt disease symptom expression was delayed, wilt disease severity was reduced, and the yield of tomatoes was enhanced. They recognized the promising antifungal activity of the nano-particles.

Xing et al. [58] prepared oleoyl-chitosan (*O*-chitosan) nano-particles (296 nm and well-dispersed nano-particles) as an antifungal dispersion system, and examined their effects in potato dextrose agar medium against *Botryosphaeria dothidea*, *Nigrospora sphaerica*, *Nigrospora oryzae*, *Alternaria tenuissima*, *Fusarium culmorum*, and *Gibberella zeae*. They observed that, except for two later fungi, others were sensitive to the antifungal compound. Furthermore, they reported unsaturated fatty acids in sensitive fungi.

Liang et al. [59] encapsulated avermectin, which is a bio-nematocide effective on pinewood nematode, through poly-γ-glutamic acid and chitosan. They observed that both nano-particles showed more mortality rates of nematodes, less photolysis, and more efficiency than avermectin. The researchers showed greater concentrations of these nano-particles in the intestines and heads of nematodes compared to avermectin. Furthermore, the release rate of avermectin was controllable under pH, and rapid release was observed in alkaline conditions.

4.4   CONCLUSIONS AND PROSPECTS

Today, in the growing world, responding to this question of how to feed the increasing population without risk or with minimum risk for human health and a healthy environment has special importance, and this is what scientists and researchers in the field of agriculture and foods are looking for. Understanding plant physiology and the inspiration of plant nature against biotic and abiotic stresses can help humans to achieve this target and sustainable agriculture and sustainability of food production, especially against biotic stress, because biotic stress can be considered to be an abiotic stress-induced doubled problem. In conclusion, it was observed that plants have both special and general mechanisms against pathogens, including biosynthesis enzymes and non-enzymatic compounds, especially biopolymers. These biopolymers are categorized in various types of materials, and proteins and polysaccharides (such as chitosan and alginates) have generalized matters. These compounds work as elicitors for the induction of plant defense systems and even more so as an antimicrobial compound as well as in plant growth and development regulation. Therefore, their exogenous application in various formulations, including poly- and oligomers, and nano-sized/nano-composites, has shown promising results. However, nanotechnology could hold greater promise because of the special characteristics of particles. Furthermore, exogenous applications of biopolymers are not limited to their direct usage; they can be used to coat bio-fertilizers such as protein elicitors and beneficial microbes to protect them and promote sustainability in agriculture. There is every prospect of interest in these bio- and eco-friendly compounds at commercial and economic levels, and each finding demands more research and development .

REFERENCES

1. Kumaraswamy RV, Kumari S, Choudhary RC et al. Engineered chitosan based nanomaterials: bioactivities, mechanisms and perspectives in plant protection and growth. International Journal of Biological Macromolecules, 2018; 113: 494–506.
2. Pour MM, Saberi-Riseh R, Mohammadinejad R, Hosseini A. Investigating the formulation of alginate-gelatin encapsulated *Pseudomonas fluorescens* (VUPF5 and T17–4 strains) for controlling *Fusarium solani* on potato. International Journal of Biological Macromolecules, 2019; 133: 603–613.
3. Choudhary MK, Rav KS, Saharan V. Novel prospective for chitosan based nano-materials in precision a trials in precision agriculture – a review. The Bioscan, 2016; 11(4): 2287–2291.
4. Sathiyabama M. Biopolymeric nanoparticles as a nanocide for crop protection. *In*: Nanoscience for Sustainable Agriculture. 2019; Springer, Cham. pp. 139–152.
5. Popletaeva SB, Statsyuk NV, Voinova TM et al. Evaluation of eliciting activity of peptidilprolyl cys/trans isomerase from *Pseudomonas fluorescens* encapsulated in sodium alginate regarding plant resistance to viral and fungal pathogens. AIMS Microbiology, 2018; 4(1): 192.
6. Holt III BF, Mackey D, Dangl JL. Recognition of pathogens by plants. Current Biology, 2000; 10(1): R5–R7.
7. Velásquez AC, Castroverde CDM, He SY. Plant–pathogen warfare under changing climate conditions. Current Biology, 2018; 28(10): R619–R634.
8. Hua C, Zhao JH, Guo HS. Trans-kingdom RNA silencing in plant–fungal pathogen interactions. Molecular Plant, 2018; 11(2): 235–244.
9. Alfano JR, Collmer A. Bacterial pathogens in plants: life up against the wall. The Plant Cell, 1996; 8(10): 1683.
10. Kaur S, Samota MK, Choudhary M et al. How do plants defend themselves against pathogens? Biochemical mechanisms and genetic interventions. Physiology and Molecular Biology of Plants, 2022; 1–20.
11. Ye J, Zhang L, Zhang X, Wu X, Fang R. Plant defense networks against insect-borne pathogens. Trends in Plant Science, 2021; 26(3): 272–287.
12. Liu J, Zhang L, Yan D. Plasmodesmata-involved battle against pathogens and potential strategies for strengthening hosts. Frontiers in Plant Science, 2021; 12: 1–14.
13. Han GZ. Origin and evolution of the plant immune system. New Phytologist, 2019; 222(1): 70–83.
14. Ramirez-Prado JS, Piquerez SJM, Bendahmane A et al. Modify the histone to win the battle: chromatin dynamics in plant–pathogen interactions. Frontiers in Plant Science, 2018; 9: 355.
15. Lopez-Moya F, Suarez-Fernandez M, Lopez-Llorca LV. Molecular mechanisms of chitosan interactions with fungi and plants. International Journal of Molecular Sciences, 2019; 20(2): 332.
16. Lee MH, Jeon HS, Kim SH et al. Lignin-based barrier restricts pathogens to the infection site and confers resistance in plants. The EMBO Journal, 2019; 38(23): e101948.
17. Yang C, Dolatabadian A, Fernando WD. The wonderful world of intrinsic and intricate immunity responses in plants against pathogens. Canadian Journal of Plant Pathology, 2022; 44(1): 1–20.
18. Zheng F, Chen L, Zhang P et al. Carbohydrate polymers exhibit great potential as effective elicitors in organic agriculture: a review. Carbohydrate Polymers, 2020; 230: 115637.

19. de Wit PJ. How plants recognize pathogens and defend themselves. Cellular and Molecular Life Sciences, 2007; 64(21): 2726–2732.
20. Beattie GA. Water relations in the interaction of foliar bacterial pathogens with plants. Annual Review of Phytopathology, 2011; 49: 533–555.
21. Bauters L, Stojilković B, Gheysen G. Pathogens pulling the strings: effectors manipulating salicylic acid and phenylpropanoid biosynthesis in plants. Molecular Plant Pathology, 2021; 22(11): 1436–1448.
22. Gullner G, Komives T, Király L, Schröder P. Glutathione S-transferase enzymes in plant–pathogen interactions. Frontiers in Plant Science, 2018; 1836.
23. Deboever E, Deleu M, Mongrand S, Lins L, Fauconnier ML. Plant–pathogen interactions: underestimated roles of phyto-oxylipins. Trends in Plant Science, 2020; 25(1): 22–34.
24. Förderer A, Yu D, Li E, Chai J. Resistosomes at the interface of pathogens and plants. Current Opinion in Plant Biology, 2022; 67: 102212.
25. Zhou Y, Yang K, Yan Q et al. Targeting of anti-microbial proteins to the hyphal surface amplifies protection of crop plants against *Phytophthora* pathogens. Molecular Plant, 2021; 14(8): 1391–1403.
26. Chen S, Zhang L, Cai X et al. (E)-Nerolidol is a volatile signal that induces defenses against insects and pathogens in tea plants. Horticulture Research, 2020; 7:
27. Díez-Pascual AM. Biopolymer composites: synthesis, properties, and applications. International Journal of Molecular Sciences, 2022; 23(4): 2257.
28. Pusztahelyi T. Chitin and chitin-related compounds in plant–fungal interactions. Mycology, 2018; 9(3): 189–201.
29. Gong BQ, Wang FZ, Li JF. Hide-and-seek: chitin-triggered plant immunity and fungal counterstrategies. Trends in Plant Science, 2020; 25(8): 805–816.
30. Romanazzi G, Feliziani E, Sivakumar D. Chitosan, a biopolymer with triple action on postharvest decay of fruit and vegetables: eliciting, antimicrobial and film-forming properties. Frontiers in Microbiology, 2018; 2745.
31. Saharan V, Pal A. Biological activities of chitosan-based nanomaterials. *In*: Chitosan Based Nanomaterials in Plant Growth and Protection. 2016; Springer, New Delhi. pp. 33–41.
32. Skorik Y, Thakur V, Tamanadar E, Noghabi S. Encapsulation of plant biocontrol bacteria with alginate as a main polymer material. International Journal of Molecular Sciences, 2021; 22(20).
33. Zhang C, Wang W, Zhao X, Wang H, Yin H. Preparation of alginate oligosaccharides and their biological activities in plants: a review. Carbohydrate Research, 2020; 494: 108056.
34. Saberi-Riseh R, Moradi-Pour M, Mohammadinejad R, Thakur VK. Biopolymers for biological control of plant pathogens: advances in microencapsulation of beneficial microorganisms. Polymers, 2021; 13(12): 1938.
35. Saberi-Riseh R, Skorik YA, Thakur VK et al. Encapsulation of plant biocontrol bacteria with alginate as a main polymer material. International Journal of Molecular Sciences, 2021; 22(20): 11165.
36. Jurić S, Đermić E, Topolovec-Pintarić S, Bedek M, Vinceković M. Physicochemical properties and release characteristics of calcium alginate microspheres loaded with *Trichoderma viride* spores. Journal of Integrative Agriculture, 2019; 18(11): 2534–2548.
37. Fathi F, SaberiRiseh R, Khodaygan P, Hosseini S, Skorik YA. Microencapsulation of a *Pseudomonas* strain (VUPF506) in alginate–whey protein–carbon nanotubes and next-generation sequencing identification of this strain. Polymers, 2021; 13(23): 4269.

38. Fathi F, Saberi-Riseh R, Khodaygan P. Survivability and controlled release of alginate-microencapsulated *Pseudomonas fluorescens* VUPF506 and their effects on biocontrol of *Rhizoctonia solani* on potato. International Journal of Biological Macromolecules, 2021; 183: 627–634.

39. Shen Y, Li J, Xiang J, Wang J, Yin K, Liu Q. Isolation and identification of a novel protein elicitor from a *Bacillus subtilis* strain BU412. AMB Express, 2019; 9(1): 1–9.

40. Aitouguinane M, Bouissil S, Mouhoub AA et al. Induction of natural defenses in tomato seedlings by using alginate and oligoalginates derivatives extracted from Moroccan brown algae. Marine Drugs, 2020; 18(10): 521.

41. Ebrahimi-Zarandi M, Skorik YA. Alginate-induced disease resistance in plants. Polymers, 2022; 14(4). https://doi.org/10.3390/polym14040661.

42. Carina D, Sharma S, Jaiswal AK, Jaiswal S. Seaweeds polysaccharides in active food packaging: a review of recent progress. Trends in Food Science & Technology, 2021; 110: 559–572.

43. Zhang C, Howlader P, Liu T et al. Alginate oligosaccharide (AOS) induced resistance to Pst DC3000 via salicylic acid-mediated signaling pathway in *Arabidopsis thaliana*. Carbohydrate Polymers, 2019; 225: 115221.

44. Arafa EG, Sabaa M W, Mohamed RR et al. Preparation of biodegradable sodium alginate/carboxymethyl chitosan hydrogels for the slow-release of urea fertilizer and their antimicrobial activity. Reactive and Functional Polymers, 2022; 174: 105243.

45. Bouissil S, Guérin C, Roche J et al. Induction of defense gene expression and the resistance of date palm to *Fusarium oxysporum* f. sp. *albedinis* in response to alginate extracted from *Bifurcaria bifurcata*. Marine Drugs, 2022; 20(2): 88.

46. Bouissil S, Alaoui-Talibi E, Pierre G et al. Use of alginate extracted from Moroccan brown algae to stimulate natural defense in date palm roots. Molecules, 2020; 25(3): 720.

47. Liu J, Kennedy JF, Zhang X et al. Preparation of alginate oligosaccharide and its effects on decay control and quality maintenance of harvested kiwifruit. Carbohydrate Polymers, 2020; 242: 116462.

48. Chen J, Wu A, Yang M et al. Characterization of sodium alginate-based films incorporated with thymol for fresh-cut apple packaging. Food Control, 2021; 126: 108063.

49. De Corato U, Salimbeni R, De Pretis A, Avella N, Patruno G. Use of alginate for extending shelf life in a lyophilized yeast-based formulate in controlling green mold disease on citrus fruit under postharvest condition. Food Packaging and Shelf Life, 2018; 15: 76–86.

50. Mahcene Z, Khelil A, Hasni S et al. Development and characterization of sodium alginate based active edible films incorporated with essential oils of some medicinal plants. International Journal of Biological Macromolecules, 2020; 145: 124–132.

51. Sun Y, Wu H, Xu S et al. Roles of the EPS66A polysaccharide from *Streptomyces* sp. in inducing tobacco resistance to tobacco mosaic virus. International Journal of Biological Macromolecules, 2022; 209: 885–894.

52. Chun SC, Chandrasekaran M. Chitosan and chitosan nanoparticles induced expression of pathogenesis-related proteins genes enhances biotic stress tolerance in tomato. International Journal of Biological Macromolecules, 2019; 125: 948–954.

53. Paul SK. Application of nanomaterials in plant protection. Seminar paper. Bangabandhu Sheikh Mujibur Rahman Agricultural University. 2020. 22p.

54. Moradi Pour M, SaberiRiseh R, Skorik YA. Sodium alginate–gelatin nanoformulations for encapsulation of *Bacillus velezensis* and their use for biological control of pistachio gummosis. Materials, 2022; 15(6): 2114.

55. Saberi-Rise R, Moradi-Pour M. The effect of *Bacillus subtilis* Vru1 encapsulated in alginate–bentonite coating enriched with titanium nanoparticles against *Rhizoctonia solani* on bean. International Journal of Biological Macromolecules, 2020; 152: 1089–1097.
56. Kanagaraj SSP, Rajaram SK, Ahamed M et al. Antimicrobial activity of green synthesized biodegradable alginate–silver (Alg–Ag) nanocomposite films against selected foodborne pathogens. Applied Nanoscience, 2021; 1–12.
57. Sathiyabama M, Charles RE. Fungal cell wall polymer based nanoparticles in protection of tomato plants from wilt disease caused by *Fusarium oxysporum* f. sp. *lycopersici*. Carbohydrate Polymers, 2015; 133: 400–407.
58. Xing K, Shen X, Zhu X et al. Synthesis and in vitro antifungal efficacy of oleoyl-chitosan nanoparticles against plant pathogenic fungi. International Journal of Biological Macromolecules, 2016; 82: 830–836.
59. Liang W, Yu A, Wang G et al. Chitosan-based nanoparticles of avermectin to control pine wood nematodes. International Journal of Biological Macromolecules, 2018; 112: 258–263.

5 Role of Enzyme-Mimicking Nanoparticles in Crop Plants

R S Pal, Tilak Mondal, Rahul Dev,
Manoj Parihar, Amit Kumar, Devender Sharma,
and Lakshmi Kant

CONTENTS

5.1 INTRODUCTION

Nanozymes are sophisticated nanomaterials with distinct physicochemical properties and the capacity to imitate genuine physiologically applicable responses through precision structural construction. Nanozymes, in particular, imitate real enzymes and act as enzyme packages. Enzymatic catalytic reactions are largely effective, with responses occurring quickly even under mild conditions, and similar responses are also largely selective. For viewing and covering operations, high efficacy and selectivity are extremely desirable components. However, natural enzymes, including proteins, have constraints such as low thermostability and a small pH window, which denature the enzymes and impair or prevent their enzymatic activity. Susceptible denaturation complicates the interpretation of sensing and monitoring data, perhaps leading to a false-positive or negative result. Nanozymes, in this context, address these limits by providing great structural durability and stability while maintaining optimal catalytic activity. Nanozymes have potential applications in a variety of sectors, including biomedicine and the environment, due to their unique physicochemical features and enzyme-like capabilities.

The current biological roles of nanozymes in plant systems as powerful catalytic tools as well as the practical deployment of these nanozymes in terms of their functionality and recyclability are important aspects. Various types of nanomaterials have been found to have intrinsic enzyme-like activity in the last few decades [1, 2]. Natural enzymes have intrinsic catalytic ability to catalyse a specific chemical change, usually at a single active site [3, 4]. Because nanozymes lack an active site, researchers have proposed a variety of ways to improve the catalytic characteristics of these nanomaterials, allowing them to selectively and effectively react with target molecules. The potential role of these enzyme-mimicking nanoparticles (NPs) in agricultural plants and a demonstration of how nanozymes have recently been used in plants are discussed in the following paragraphs.

5.2 TYPES OF NANOZYMES

Nanozymes, which mimic natural enzymes, exhibit enzyme-like properties but are more active and overcome the limitations of natural enzymes, such as low thermal stability and narrow pH window, which will denature the enzymes, greatly reducing and inhibiting their enzymatic activities. They are categorized into four types based on their mode of natural enzyme-mimicking behaviour (Figure 5.1) [5–8]. Carbon-based nanozymes have been synthesized and very easily modified. They have been extensively considered in many different applications due to their specific electronic and geometric characters that facilitate mimicking the catalytic sites of natural enzymes [9,

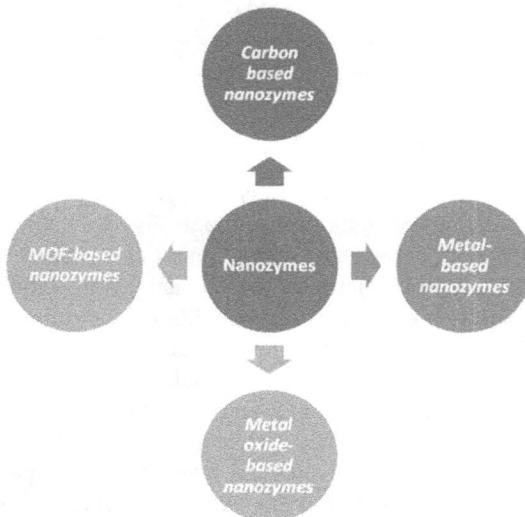

FIGURE 5.1   Types of nanozymes. MOF, metal-organic framework.

10]. They have revealed strong functional stability and exceptional robustness against severe situations. Examples of important carbon-based nanozymes are fullerenes, carbon nanotubes (CNTs), graphene, graphene oxide, carbon dots, graphene quantum dots and carbon nitride, which is used to mimic the catalytic activities of superoxide dismutase, peroxidase, oxidase, hydrolase and catalase enzymes [11].

Metal-based nanozymes are synthesized from metals that are known to exhibit intrinsic catalytic behaviour for various heterogeneous reactions that catalyse the same reactions as natural enzymes. Such metals include gold, silver, platinum, palladium and iridium. Metal-based nanozymes have shown their specific electronic and catalytic properties due to this property; these types of nanozymes have been extensively explored as artificial enzymes. They were found to exhibit different natural enzyme-like activities, including oxidase, peroxidase, catalase, glucose oxidase and superoxide dismutase [12–15].

Metal oxide-based nanozymes utilize metal oxide and metal sulphide to mimic the metal catalytic active site found in the metal–hemeredox centre of metallo-enzymes. The most common example of such types of nanozymes is iron oxide (Fe_3O_4). Cerium oxide (CeO_2) NPs were used due to their unique ability to interchange between valence states, which exhibit peroxidase, multifunctional catalase and superoxidase-like activity [16–18].

Metal-organic frameworks (MOFs) are comprised between the coordinate bonds of organic ligands and metal ions which were organized in a periodic structure, with specified lengths and highly ordered bonds, leading to a porous or channel-like morphology [19–21]. Due to the high surface area originating from these cavities and the possibility of pore size control through synthesis and physicochemical modifications, MOFs have gained increased popularity for several applications. MOFs can mimic the function of varied enzymes, including superoxide dismutase, catalase, methane monooxygenase, oxidase, peroxidase and lipase.

5.2.1 CATALYTIC MECHANISMS OF NANOZYMES

The catalytic mechanisms of natural enzymatic activities influence the fundamental functions of nanozymes in the system. The catalytic activities of nanozymes depend on their size, shape and composition. The mode of catalytic reactions of natural enzymes involves chemical and binding mechanisms [7, 8]. Chemical mechanisms include acid–base, covalent and metal ion catalysis, whereas the binding mechanisms of natural enzymes involved proximity/orientation-assisted catalysis, transition state stabilization-assisted catalysis and electrostatic catalysis. Nanozymes also employ one or more of these mechanisms to mimic natural enzymes. The catalytic mechanism of carbon-based nanozymes is not as well documented; carbon-based nanozymes (graphene oxide quantum dots) mimic the mechanism of the peroxidase-like activity which can catalyse the decomposition of hydrogen peroxide and generate highly reactive hydroxyl radicals. These catalytic reactions occur at reactive and substrate-binding sites. The functional groups –C=O act as catalytically active sites for converting hydrogen peroxide to hydroxyl radicals, and the O=C–O– groups serve as the substrate-binding sites for hydrogen peroxide [10, 11]. The mechanism of metal-based nanozymes depends on catalysis of the redox reaction, and this generally includes the procedure of transition metal elements which have variable oxidation states and change their valance. The metal-based nanozymes exhibit intrinsic pH-switchable peroxidase and catalase-like activities [7–9], which consist of the adsorption and decomposition of hydrogen peroxide under different pH conditions. For metal-based nanozymes, the catalytic mechanism arises from the adsorption, activation and electron transfer of the substrate on to the metal surface, in contrast to the mechanisms occurring by changes in the metal valence of the nanomaterial, as in the case of other metal compound-based nanozymes [4–7]. Different works have reported peroxidase-like activities of Au, Ag, Pt and Pd nanomaterials at acidic pH, and catalase-like activities at basic pH. The peroxidase-like activities at low pH are attributed to the base-like decompositions of H_2O_2 on the surfaces of metal, and at high-pH conditions the catalase-like activities are attributed to the acid-like decompositions of H_2O_2 [12–15]. The metal oxide nanozyme (Fe_3O_4, CeO_2 and MnO_2) nanostructures demonstrate their activity due to their unique ability to interchange between valence states. They have been shown to mimic different kinds of enzymes, including peroxidases, oxidases, hydrolases and catalases. Iron oxide (Fe_3O_4) NPs as an example of metal-based nanozyme reported peroxidase-like activity, which originated mainly from the interaction of hydrogen peroxide with ferrous ions on the surface of the NPs.

Iron oxide nanostructures have also been used for the Fenton reaction between the Fe^{2+} sites and the adsorbed H_2O_2. This reaction involves the initial formation of a Fe^{2+}–H_2O_2 complex, which produces •OH radicals and other reactive oxygen species (like HO_2• and O_2•–), and a radical chain reaction is promoted [16, 17]. The catalytic reactions of MOF nanozymes have not yet been well described; there are so many theories or explanations available. Among the theories, the most acceptable catalytic behaviour of MOFs has been described as nanozymes, in which a reactive catalytic site originated from two characteristic features, one from the metal ion portion that can act as biomimetic catalyst centre and another from the organic ligand portion that can act as electron mediators which used to accept electrons from a substrate and donate them to another, thus catalysing reactions similarly to natural enzymes, such as superoxide dismutase, catalase, methane monooxygenase, oxidase, peroxidase and lipase [19, 20]. In addition, MOFs have special characteristics of highly specific surface area due to their cavity, pore sizes and highly exposed active sites which have contributed to the achievement of performances of high catalytic efficiency, and the organic ligands can supply suitable optical and electrical characteristics with abundant functional groups for chemical modifications.

Apart from the different theories of the catalytic mechanism of different nanozymes, the catalytic performance of nanozymes has been determined by the Michaelis–Menten model which is governed by steady-state kinetic parameters, such as the Michaelis constant (K_m) and maximal reaction rate (V_{max}), which are commonly employed to evaluate the performance of nanozymes [21]. V_{max} expresses the reactivity of the nanozyme when saturated with the substrate at a fixed concentration. K_m is described as the substrate concentration when the enzymatic reaction reaches half of V_{max}. According to this model, K_m value is inversely proportional to the affinity of the nanozyme to the substrate, i.e. a small K_m indicates a high affinity.

5.3 NANOZYMES IN PLANT STRESS TOLERANCE

Agriculture production and extension are severely hampered by abiotic stressors. Drought, salinity, alkalinity, submergence, mineral and metal toxicity/deficiencies and a variety of other factors all inhibit crop development and productivity. To battle various pressures, plants adapt and alleviate abiotic stresses by changing their morphological, physiological, biochemical and molecular structures. Nanotechnology has enormous potential to modify conventional plant systems [22]. Chemically synthesized agrochemicals, such as fungicides and pesticides, are sprayed to prevent microbial crop diseases. But these chemicals are quite hazardous to plants and are potentially toxic to human health [23]. In most cases, agrochemicals applied in fields are unable to reach the targeted sites due to factors like leaching, hydrolysis, photolysis and especially microbial degradation [24]. Due to nano-size, nano-fertilizers and nano-pesticides can be easily distributed in a controlled fashion with precise specificity, thereby reducing collateral damage. Nanotechnological applications in farming have gained attention because of their well-organized control and accurate release of herbicides, pesticides and fertilizer [25]. The role of agrochemicals is crucial in modern agriculture, but the development of nano-fertilizers and nano-pesticides transformed the agricultural sector [26]. Nanomaterials with enzyme-like activities (nanozymes) have shown the potential to augment the inherent functions of plants, e.g. photosynthesis and stress resistance. The use of the antioxidant capacities of nanozymes to augment photosynthesis and enhance stress resistance of plants has been explored recently.

It has been recognized that abiotic stress, caused by salt, drought, heat and cold, for example, results in the reduction of plant growth and loss of yields worldwide [27]. Application of antioxidant enzyme-mimicking NPs to alleviate abiotic stress and maintain crop production is a promising strategy in nano-enabled agriculture, given increased population-induced food safety issues, shrinking arable land and climate change. Utilizing the inherent properties of nanomaterials to promote plant growth and increase biomass is a promising strategy in nano-enabled agriculture. Nanozymes with enzyme-like activities have potential due to their ability to replace specific enzymes in enzyme-based applications. NPs, with remarkable multi-antioxidant enzyme-mimicking activities (reactive oxygen species (ROS)-scavenging capacity), enhanced photosynthesis and biomass accumulation in cucumber plants at an appropriate dose (1 mg per plant) through foliar application [28]. Foliar application of NPs triggered the up-regulation of antioxidant low-molecular-weight metabolites, which may stimulate the immune system and protect the plant from pathogen attack. Results encourage us to advocate that nanozymes may play important roles in plant stress resistance and have potential applications in nano-enabled agriculture, due to their unique physiochemical and catalytic activities. However, the fate and transport of NPs within a plant, and the mechanism of how nanozymes boost endogenous antioxidants, remain unknown and further studies are needed to elucidate the underlying mechanism. The prosperity of nanotechnology and biology created a series of novel artificial enzymes. As promising natural enzymes mimics, nanozymes have demonstrated remarkable performance in plant stress tolerance.

NPs enter cells either through penetration or through transit through specific channels in the cellular membrane. NPs may act as stress-signalling molecules, causing an increase in the expression of numerous genes involved in the stressed state. This involves the induction of regulatory factor expression, which leads to the activation of the defence system and, finally, stress tolerance. Aside from a tolerable level, NPs can keep ROS at a high enough level to activate the plant's defence mechanism by inducing the ROS-signalling network. The root architecture alteration after exposure to NPs, for example, could be attributable to the down-regulation of genes involved in trichoblast development. Trichoblasts fall under the category of specialized epidermal cells because this is where root hairs form. In addition, genes responsive to indole acetic acid and ethylene have been identified as positive regulators of root hair formation [29]. Nanozymes typically change cellular pathways involved in defence systems. NADPH oxidase, GST, superoxide dismutase, and peroxidases are among the genes whose expression is up-regulated by NP treatment [29]. Root architecture modification, antioxidant mechanism activation and involvement of specific phytohormone signalling pathways were found to be common NPs that induced stress responses, though the effects were influenced by the NPs' nature and duration of exposure [29, 30].

Researchers determined that NPs aid plant survival by influencing plant growth and development in a concentration-dependent manner [31]. Recapitulation of the potential interaction between NPs and plant metabolisms is required to investigate novel insights into plant stress tolerance.

5.4   NANOZYMES FOR ENHANCING PHOTOSYNTHETIC EFFICIENCY

The physiological and biochemical parameters of crop plants improved significantly after using nano-fertilizers. Nano-fertilizers are vital in current agriculture because they have the right formulations and delivery mechanisms to ensure that plants get the most out of them [32]. Nanoscale particles are smaller and can absorb with different dynamics than bulk particles or ionic salts, which has substantial advantages [33]. Nano-enabled fertilizers have shown a rise in productivity by providing targeted delivery/gradual release of nutrients and lowering fertilizer application with an increase in nutrient use efficiency [34]. The surface–mass ratio of nano-fertilizers is increased as their size is lowered through physical/chemical processes, allowing for increased nutrient absorption by the roots. The beneficial effect on total chlorophyll content in sunflower leaves was boosted by a biocompatible magnetic nano-fluid [35]. Foliar spraying of $nTiO_2$ increased photosynthetic pigments in *Zea mays*, and this was linked to increased crop output and anthocyanin and photosynthetic pigments, rubisco activity and photosynthetic efficiency in barley. The application of $nTiO_2$ in spinach improved plant performance by up to 17-fold and increased nitrogen metabolism, protein levels and green pigments by around 29% [36].

In cotton and soybean crops, nano-Zn fertilizer reduced peroxidase, catalase and oxidase enzyme activities while increasing polyphenol content. Plants produce antioxidants as secondary metabolites in response to adverse conditions such as drought, salt and nutritional shortage [37]. The nanoertilizers (NFs) provide enough nutrients to boost antioxidant activity in plant cells, with increased photo-assimilation capacity and grain yield. The use of $nTiO_2$ increased plant fresh and dry mass by improving photosynthetic capacity and nitrogen metabolism by improving pigment formation and the conversion of light energy into biochemical energy via improved photophosphorylation, which also up-regulated biological carbon sequestration through the Calvin cycle in over 95% of plants. *Zea mays* biomass and productivity were increased by the photocatalytic activity of $nTiO_2$ in nanoform [36]. The morphological properties of nanocarriers may influence nutrient transport via the surface of membranes, which is critical in demonstrating the applicability and utility of nano-fertilizers. In chickpea, maize and tomato seedlings, it was discovered that nanochitosan has a positive influence on morphological and physiological aspects in both germinating seed and foliar treatment to improve seedling growth, biomass, germination ability and seed vigour index [38]. NPs are responsible for the loss of auxins, cytokinins and salicylic acid, signifying a hormonal imbalance in plants that impacts general metabolism [39]. In order to face environmental adversities for survival, zinc may stimulate important enzymes associated with biochemical processes, such as glucose and protein growth regulator metabolism, pollen production and membrane integrity [40], as well as terminal oxidase in mitochondria.

5.5 NANOSENSORS FOR STRESS SENSING

Under stress conditions, crop plants develop complex mechanisms, including sensing and production of various signalling molecules. These molecules include H_2O_2, sugars (such as glucose and sucrose), Ca^{2+}, gas molecules (such as nitric oxide, hydrogen sulphide and carbon monoxide) and plant hormones such as ethylene, jasmonic acid, abscisic acid, methyl salicylate and volatile organic compounds [41, 42]. The role of these signalling molecules, determined by the extent of stress, such as Ca^{2+} role, remains vital under salinity and osmotic stress and produces various peaks [43]. However, real-time detection of these signalling molecules is limited and more studies should be conducted for detailed investigation. To monitor the role of signalling molecules, plant biotechnology could be an effective tool for non-model crop plants. For example, to detect glucose molecules ratiometric quantum dot [44], for H_2O_2, hemin-complexed DNA aptamer-coated single-walled CNTs [45], for nitric oxide (NO), AT15-coated CNTs [46, 47] and for Ca^{2+} nanoneedle transistor-based sensors could be employed. Moreover, to sense the climatic variables such as temperature [48], humidity [49] and stomatal activities [50], nano-sensors have been developed. Recently, researchers developed NP-based conducting ink for the real-time study of opening and closing of stomata [50].

In the field of nano-sensors, our understanding of plant mechanisms to sense Na^+ under stress conditions and the development of Na^+-specific sensors to detect Na^+ transportation with greater resolution must be developed. Similarly, tools to define the roles of hydroxyl species, as an important ROS under stress conditions, require proper attention and thorough investigation [42]. Current dye-based visualization methods including hydroxyphenyl fluorescein are not solely developed for sensing hydroxyl species [51], which limits our understanding to study their role and importance in plant development, their growth and under stress alleviation. In this regard, development of hydroxyl-specific nano-sensors would provide greater help to investigate their function in plant systems.

5.6 MAKING PLANT SENSORS FOR EARLY STRESS DETECTION

Early stress detection technology could prove to be a revolutionary technology in improving agriculture production. At present, we use remote sensing or hyperspectral imaging to monitor plant water stress, chlorophyll content, morphological changes, etc. However, plants exhibit these traits or changes after accumulation of stress, which results in significant losses in plant productivity [47]. In this condition, early detection using nano-sensors provides greater spatial and temporal resolution and facilitates remote sensing technology to monitor stress signalling molecules and thereby stress conditions in plants. Nano-sensors are able to track and monitor stress-signalling molecules in plants, such as glucose, H_2O_2 and NO, and allow transduction of chemical signals to optical or radio waves, which are further detected by agricultural equipment. Such tools facilitate better farm management and early detection of stress conditions in agricultural crops. Recently, nano-enabled plants have been demonstrated to be potential engineering smart sensors to identify or detect early stress conditions. However, these nano-enabled plants are still in their nascent phase and this technology must be refined under real-time field conditions. Under field conditions, several stresses, such as temperature, drought, salinity, occur in combination [52]. Multiple stresses complicate the chemical signalling process and under such condition signals received from nano-sensors must be of higher resolution in response to various signalling molecules. Nano-sensors with greater accuracy and sensitivity would enable higher-resolution signals to be decoded under real-time field conditions and thus their application in agricultural field would be increased. Efforts should be made to identify new nano-sensors for signalling molecules which are currently not available and to prepare a research database of the behaviour of various signalling molecules, such as Ca^{2+} signatures and ROS signatures under stress conditions. In addition, use of ZnO NPs at room temperature under organic field allowed signalling molecules to be sensed, such as carbon monoxide, for monitoring early plant stress conditions [53].

5.7 NANO-ENABLED TRANSGENIC PLANTS

Around 10,000 years ago, farming began. Farmers utilized basic selection followed by hybridization before genetic engineering became available. More sophisticated procedures such as somatic hybridization (cell fusion), somaclonal variation and mutant breeding were created later on as biotechnologies improved. Genetic engineering began in 1973; recently, NPs have proven to be a game-changing instrument for addressing a wide range of global concerns, including those affecting agriculture. An NP is defined as any particle with a typical size of less than 100 nm. Nanotechnology has played a key role in the fields of genetic engineering and plant transformation, making it a promising choice for plant optimization and manipulation. Most genetic modifications to plants were previously carried out using

Agrobacterium or tools such as the gene gun (biolistics); however, these traditional methods of gene transformation face barriers due to low species compatibility/genotype dependency, lack of versatility/compatibility with chloroplastial/mitochondrial gene transformations, low transformation efficiency and the risk of damage to plant tissues (cell or organelle damage) due to the impact of biolistics. Traditional techniques of plant transformation (*Agrobacterium* and biolistics) risk DNA incorporation in the plant genome, rendering it transgenic and certifying it as a genetically modified organism. Non-incorporative/DNA-free genetic alterations have become a highly significant topic of study [54]. CNT- and different porous NP-enabled delivery techniques are used in nano-enabled technologies, which may allow for higher-throughput plant alteration while avoiding legal genetically modified organism constraints [55]. For the transfer of genetic material, a revolutionary technique employs highly tailorable diffusion-based nanocarriers, allowing for non-transgenic, non-destructive crop improvement. Through the use of nanocarriers, nanomaterial-mediated gene delivery ensures the proper protection of transferred DNA from the nucleases present in the cell cytoplasm. The issue of genetic manipulation in some valued refractory plant genotypes can be resolved by conjugating desired nucleic acids with designed nanocarriers.

5.7.1 Factors Affecting Gene Transformations

The method's specificity is largely reliant on the material's characteristics, with size, polarity and surface chemistry being critical considerations. The multilayered and rigid cell wall is the major barrier to successful and efficient gene delivery to plant cells. In comparison to more traditional transformation techniques like biolistics and/or *Agrobacterium*, the high cost and complicated production of certain nanocarriers may be the principal barriers to large-scale usage of NPs in plant transformation investigations [56].

5.7.2 Nanosystem Delivery Methods

The nanocarrier/vesicles, which can transport the macromolecule into cells, should readily pass through the cell wall to reach the plasma membrane in order to target the nucleus and cytoplasm to create an effective nanomaterial-based gene delivery system [57]. There are seven significant categories of NPs, including dendrimer NPs, nanocomposites, nanofibres and nanocapsules [58]. The four main types of NPs used for gene delivery that are the most popular [59, 60] are:

1. Carbon-based NPs
2. Polymeric-based NPs
3. Metallic/magnetic NPs
4. Polymeric-based NPs.

5.7.3 Vesicles of Genetic Transformation

Nano-structured DNA [61], single-walled CNTs [62], liposomes, organic polymers and metals or metal oxides (i.e., silver (Ag), zinc oxide (ZnO) and titanium dioxide (TiO_2)), silica (SiO_2), layered double hydroxides (LDH) [63, 64] and other NPs [65] have been employed as vesicles for the transfer of genetic information into chloroplasts and nuclei in several diffusion-based delivery techniques (Figure 5.2).

Porous NPs have been found to be an efficient DNA delivery method for plant transformation, with pore size and strand length influencing efficacy [65]. Overall, diffusion-based gene transformation approaches are a less expensive way of plant gene transformation that has a smaller impact on plant tissue, lower transformation efficiencies and little to no danger of DNA incorporation. Because the particle being released into the cell is dramatically smaller, the effect may be lowered while maintaining a similar level of genetic transformation efficiency as classical biolistics.

However, because the majority of studies involving nanoscale biolistic techniques are conducted on animal cells, plant transformation is still relatively new and may face hurdles not observed in animal cell studies [66]. Plant-based nano-enabled genetic engineering, on the other hand, lags behind animal-based genetic engineering, owing to a number of factors. Plant cells, unlike animal cells, have a wall enclosing cell membrane to offer mechanical and structural support. The existence of this plant cell wall, which enables only biomolecules with a diameter of less than 20 nm to get through, limits the use of nanomaterials in genetically modified crops. Therefore, NPs, for example, quantum dots, NPs, nanotubes, liposomes and DNA nanostructures, must be smaller than 20 nm in at least one dimension to be employed for plant transformation [67]. Breakthroughs in nanotechnology for genetic engineering have made it easier to change the genetic makeup of plants in recent years. For example, scientists have devised a method for delivering genes into the chloroplasts

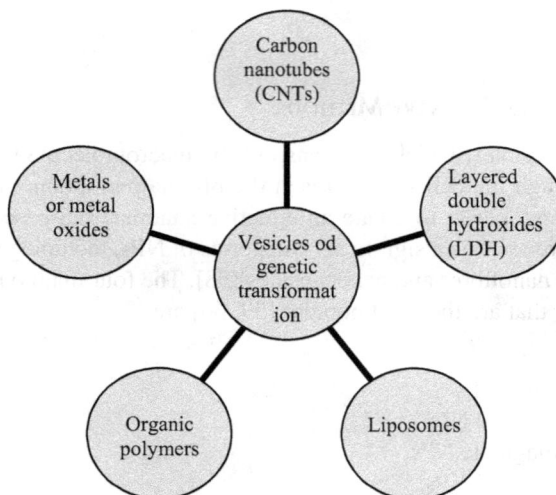

FIGURE 5.2 Nanocarrier/vesicles of genetic transformation.

of plant cells that works with a variety of plant species, including spinach and other crops [68].

5.7.4 ADVANTAGES OF NANO-ENABLED GENETIC TRANSFORMATION

The main differences between nanobiotechnology and traditional approaches for genetic transformation are speed, accuracy, dependability and scope rather than delivery objectives (species, cell and tissue-specificity) [56]. Nanomaterial-mediated plant genetic transformation provides the following benefits over typical plant transformation methods:

- It can pass through the plant cell wall without the use of a power source.
- By loading a huge number of nucleic acids, it is capable of achieving enormous gene fragments and multigenetic transformation.
- It is not limited to a single host.
- It has the ability to protect external nucleic acids and lower the rate of nucleic acid breakdown in cells.
- The efficiency of genetic transformation is greater compared to traditional methods [67].

Nanotechnology, in general, offers an innovative and competitive method for plant genetic change. Future study into the uses of these technologies will focus on a wider range of crops, with the goal of using less expensive, more scalable ways and investigating potential environmental consequences. Finally, whether NP plant transformations will become a common practice in the future of agriculture will be determined by these design requirements.

5.8 CONCLUSION AND FUTURE PERSPECTIVES

Currently, nanozymes have significant relevance owing to their low production cost, quick large-scale production and higher catalytic stability. Nanozymes can be modified and purified very easily compared to naturally occurring enzymes. In modern nano-enabled agriculture utilization of innate behaviour of nanozymes/NPs is a very promising strategy. This review clearly describes the prospects of nanozymes in the growth and development of crop plants. The maintenance and/or improvement of crop plant resistance against diverse stresses – biotic as well as abiotic –is key to sustainable agriculture and food security. The engineered NPs improve seed germination, plant growth and development, and also help in pathogen diagnostics and toxic agrochemicals detection. Though destruction of ROS, synergy in nitrogen fixation, stress sensing and early detection, targeted delivery and controlled release of agrochemicals, nanozymes or nanomaterials have the potential to improve plant performance. Investigations have shown that optimum foliar application (1 mg/plant) of Mn_3O_4 NPs in cucumber also activates endogenous antioxidant defence pathways as well as working as an exogenous antioxidant. However, overdose and exposure to nanomaterials might induce ROS overproduction and lead to oxidative stress. Boosting the nutritional status of crop plants through NPs not only improves yield and nutrient use efficiency of crops but also facilitates pathogen suppression, and hence can play dual roles as nano-fertilizers and nano-pesticides. Nanozymes have extensive uses due to their distinguished functions and capacities in biosensing and immunoassay, which is the main reason for designing and engineering nanozymes. NP-based CRISPR-Cas delivery has recently demonstrated benefits over previous approaches, including enhanced stability, minimal toxicity, high loading capacity and a wide range of recipient plant species, among others. However, it is currently uncertain whether NPs can facilitate CRISPR-Cas genome editing in plant cells or organelles. Plant genetic engineering will go further with the use of nanomaterials to introduce CRISPR-Cas elements into plants. Furthermore, more research on the cytotoxicity of nanomaterials is required. For the rational design of nanomaterials for agricultural use, it is pertinent to discover the main mechanisms involved in plant environmental stress tolerance and performance. There is also a need to develop novel techniques to identify, quantify and transform nanomaterials in the crop system. It will be very pertinent to focus on the green synthesis of nanoparticles with utilization of diverse agro-industrial byproducts in order to curtail direct transfer of hazardous chemicals to the agro-ecosystem. The development of a sound regulatory framework for the mitigation of biosafety risks associated with nanomaterial use in agriculture is also required to address various concerns.

REFERENCES

1. Jiang DW, et al. Nanozyme: new horizons for responsive biomedical applications. Chemical Society Reviews, 2019; 48: 3683–3704.
2. Wang QQ, et al. Nanozyme: an emerging alternative to natural enzyme for biosensing and immunoassay. Trends in Analytical Chemistry, 2018; 105: 218–224.
3. Huang Y, Ren J, Qu X. Nanozymes: classification, catalytic mechanisms, activity regulation, and applications. Chemical Reviews, 2019; 119: 4357–4412.
4. Wang XY, Hu YH, Wei H. Nanozymes in bionanotechnology: from sensing to therapeutics and beyond. Inorganic Chemistry Frontiers, 2016; 3: 41–60.
5. Elicia LS, et al. Nanozymes for environmental pollutant monitoring and remediation. Sensors, 2021; 21: 408.
6. Ballesteros CAS, et al. Recent trends in nanozymes design: from materials and structures to environmental applications. Materials Chemistry Frontiers, 2021; 5: 7419–7451. DOI: 10.1039/D1QM00947H
7. Wang H, Wan K, Shi X. Recent advances in nanozyme research. Advanced Materials, 2019; 31: 1–10.
8. Zhang R, Yan X, Fan K. Nanozymes inspired by natural enzymes. Accounts of Materials Research, 2021; 2: 534–547.
9. Sun H, Ren J, Qu X. Nanozymology. Springer Singapore, Singapore, 2020; pp. 171–193.
10. Ding H, et al. Carbon-based nanozymes for biomedical applications. Nano Research, 2021; 14: 570–583.
11. Sun H, et al. Carbon nanozymes: enzymatic properties, catalytic mechanism, and applications. Angewandte Chemie – International Edition, 2018; 57: 9224–9237.
12. Chen W, et al., Metal and metal-oxide nanozymes: bioenzymatic characteristics, catalytic mechanism, and eco-environmental applications. Nanoscale, 2019; 11: 15783–15793.

13. Logan N, et al. Amalgamated gold-nanoalloys with enhanced catalytic activity for the detection of mercury ions (Hg^{2+}) in seawater samples. Nano Research, 2020; 13: 989–998.
14. Chen YS, et al. Fluorescence quenchers manipulate the peroxidase-like activity of gold-based nanomaterials. ACS Omega, 2020; 5: 24487–24494.
15. Landry MJ, et al. Surface-plasmon-mediated hydrogenation of carbonylscatalyzed by silver nanocubes under visible light. ACS Catalysis, 2017; 7: 6128–6133.
16. Xiao M, Li N, Lv S. Iron oxide magnetic nanoparticles exhibiting zymolyase like lytic activity. Chemical Engineering Journal, 2020; 3: 94–125.
17. Ahmed SR, Cirone J, Chen A. Fluorescent Fe_3O_4 quantum dots for H_2O_2 detection. ACS Applied Nano Materials, 2019; 2: 2076–2085.
18. Zhu M, et al. Synergistic effects between polyvinylpyrrolidone and oxygen vacancies on improving the oxidase-mimetic activity of flower-like CeO_2 nanozymes. Nanoscale, 2020; 12: 19104–19111.
19. Youni SA, et al. Rare earth metal–organic frameworks (RE-MOFs): synthesis, properties, and biomedical applications. Coordination Chemistry Reviews, 2021; 429: 213620.
20. Niu X, et al. Metalorganic framework based nanozymes: promising materials for biochemical analysis. Chemical Communications, 2020; 56: 11338–11353.
21. Huang Y, Ren J, Qu X. Nanozymes: classification, catalytic mechanisms, activity regulation and applications. Chemical Reviews, 2019; 119: 4357–4412.
22. Gao Y, Zhou YZ, Chandrawati R. Metal and metal oxide nanoparticles to enhance the performance of enzyme-linked immunosorbent assay (ELISA). ACS Applied Nano Materials, 2020; 3: 1–21.
23. Gao L, et al. Intrinsic peroxidase-like activity of ferromagnetic nanoparticles. Nature Nanotechnology, 2007; 2: 577–583.
24. Gao ZQ, et al. Magnetic bead-based reverse colorimetric immunoassay strategy for sensing biomolecules. Analytical Chemistry, 2013; 85: 6945–6952.
25. Wang ZF, et al. Fluorescent artificial enzyme-linked immunoassay system based on Pd/C nanocatalyst and fluorescent chemodosimeter. Analytical Chemistry, 2013; 85: 11602–11609.
26. Dong JL, et al. Co_3O_4 nanoparticles with multi-enzyme activities and their application in immunohistochemical assay. ACS Applied Materials & Interfaces, 2014; 6: 1959–1970.
27. Gao ZQ, et al. Urchin-like (gold core)@(platinum shell) nanohybrids: a highly efficient peroxidase-mimetic system for in situ amplified colorimetric immunoassay. Biosensors and Bioelectronics, 2015; 70: 194–201.
28. Huo MF, et al. Tumor-selective catalytic nanomedicine by nanocatalyst delivery. Nature Communications, 2017; 8: 357.
29. Wang GL, et al. Dual responsive enzyme mimicking activity of AgX (X=Cl, Br, I) nanoparticles and its application for cancer cell detection. ACS Applied Materials & Interfaces, 2014; 6: 6434–6442.
30. Maji SK, Et al. Cancer cell detection and therapeutics using peroxidase-active nanohybrid of gold nanoparticle-loaded mesoporous silica-coated graphene. ACS Applied Materials & Interfaces, 2015; 7: 9807–9816.
31. Tian ZM, et al. Highly sensitive and robust peroxidase-like activity of porous nanorods of ceria and their application for breast cancer detection. Biomaterials. 2015; 59: 116–124.
32. Asati A, et al. pH-tunable oxidase-like activity of cerium oxide nanoparticles achieving sensitive fluorigenic detection of cancer biomarkers at neutral pH. Analytical Chemistry, 2011; 83: 2547–2553.

33. Chugh G, Siddique K, Solaiman Z. Nanobiotechnology for agriculture: smart technology for combating nutrient deficiencies with nanotoxicity challenges. Sustainability, 2021; 13: 1781.

34. Kah M, Tufenkji N, White JC. Nano-enabled strategies to enhance crop nutrition and protection. Nature Nanotechnology, 2019; 14: 532–540.

35. Pirvulescua A, Salaa F, Boldea M. Variation of chlorophyll content in sunflower under the influence of magnetic nanofluids. In: Proceedings of the International Conference of Numerical Analysis and Applied Mathematics 2014 (ICNAAM-2014), 22–28 September. Rhodes, Greece.

36. Morteza E, et al. Study of photosynthetic pigments changes of maize (*Zea mays* L.) under nano TiO_2 spraying at various growth stages. Springer Plus, 2013; 2: 247.

37. Rezaei M, Abbasi H. Foliar application of nanochelate and non-nanochelate of zinc on plant resistance physiological processes in cotton (*Gossypium hirsutum* L.). Iranian Journal of Plant Physiology, 2014; 4: 1137–1144.

38. Kondal R, et al. Chitosan-urea nanocomposite for improved fertilizer applications: the effect on the soil enzymatic activities and microflora dynamics in N cycle of potatoes (*Solanum tuberosum* L.). Polymers, 2021; 13: 2887.

39. Vankova R, et al. ZnO nanoparticle effects on hormonal pools in *Arabidopsis thaliana*. Science of the Total Environment, 2017; 593: 535–542.

40. Salachna P, et al. Zinc oxide nanoparticles enhanced biomass and zinc content and induced changes in biological properties of red *Perilla frutescens*. Materials, 2021; 14: 6182.

41. Gilroy S, et al. ROS, calcium, and electric signals: key mediators of rapid systemic signaling in plants. Plant Physiology, 2016; 171: 1606–1615.

42. Mittler R. ROS are good. Trends in Plant Science, 2017; 2: 211–219.

43. Shabala S, Wu H, Bose J. Salt stress sensing and early signalling events in plant roots: current knowledge and hypothesis. Plant Science, 2015; 241: 109–119.

44. Li J, et al. Standoff optical glucose sensing in photosynthetic organisms by a quantum dot fluorescent probe, ACS Applied Materials & Interfaces, 2018; 10: 28279–28289.

45. Wu H, et al. Monitoring plant health with near-infrared fluorescent H_2O_2 nanosensors. Nano Letters, 2020; 20: 2432–2442.

46. Zhang J, et al. Single molecule detection of nitric oxide enabled by d(AT)15 DNA adsorbed to near infrared fluorescent single walled carbon nanotubes. Journal of the American Chemical Society, 2011; 133: 567–581.

47. Giraldo JP, et al. Nanobiotechnology approaches for engineering smart plant sensors. Nature Nanotechnology, 2019; 14: 541–553.

48. Giacomo RD, Daraio C, Maresca B. Plant nanobionic materials with a giant temperature response mediated by pectin-Ca^{2+}. Proceedings of the National Academy of Sciences (PNAS), USA, 2015; 112: 4541–4545.

49. Oren S, et al. High-resolution patterning and transferring of graphene-based nanomaterials onto tape toward roll-to-roll production of tape-based wearable sensors. Advanced Materials Technologies, 2017; 2: 170–223.

50. Koman VB, et al. Persistent drought monitoring using a microfluidic-printed electro-mechanical sensor of stomata: in planta. Lab Chip, 2017; 17: 4015–4024.

51. Ken-ichi Setsukinai Y, et al. Development of novel fluorescence probes that can reliably detect reactive oxygen species and distinguish specific species. Journal of Biological Chemistry, 2003; 278: 3170–3175.

52. Suzuki N, et al. Abiotic and biotic stress combinations. New Phytologist, 2014; 203: 32–43.

53. Narayana A, et al. Green and low-cost synthesis of zinc oxide nanoparticles and their application in transistor-based carbon monoxide sensing. RSC Advances, 2020; 10: 13532–13542.
54. Demirer GS, et al. High aspect ratio nanomaterials enable delivery of functional genetic material without DNA integration in mature plants. Nature Nanotechnologist, 2019; 14: 456–464.
55. Wang JW, et al. Nanoparticle-mediated genetic engineering of plants. Molecular Plant Cell Press, 2019; 12(8): 1037–1040.
56. Niazian M, et al. Perspectives on new opportunities for nano-enabled strategies for gene delivery to plants using nanoporous materials. Planta, 2021; 254(4): 1–20.
57. Torney F, et al. Mesoporous silica nanoparticles deliver DNA and chemicals into plants. Nature Nanotechnology, 2007; 2: 295–300.
58. Kanwar MK, et al. Impacts of metal and metal oxide nanoparticles on plant growth and productivity. *In*: Nanomaterials and Plant Potential. Husen A, Iqbal M (eds.). 2019; Springer, Cham. pp. 379–392.
59. Sandhu KK, et al. Gold nanoparticle-mediated transfection of mammalian cells. Bioconjugate Chemistry, 2002; 13: 3–6.
60. Zhao P, et al. Synthesis and drug delivery applications for mesoporous silica nanoparticles. Avicenna Journal of Medical Biotechnology, 2017; 1: 1–8.
61. Zhang H, et al. DNA nanostructures coordinate gene silencing in mature plants. Proceedings of the National Academy of Sciences of the USA, 2019; 116(15): 538–678.
62. Demirer GS, et al. High aspect ratio nanomaterials enable delivery of functional genetic material without DNA integration in mature plants. Nature Nanotechnology, 2019; 14(5): 456–464.
63. Gogos A, Knauer K, Bucheli TD. Nanomaterials in plant protection and fertilization: current state, foreseen applications, and research priorities. Journal of Agricultural and Food Chemistry, 2012; 60(39): 9781–9792.
64. de Oliveira JL, et al. Application of nanotechnology for the encapsulation of botanical insecticides for sustainable agriculture: prospects and promises. Biotechnology Advances, 2014; 32(8): 1550–1561.
65. Hussain HI, et al. Mesoporous silica nanoparticles as a biomolecule delivery vehicle in plants. Journal of Nanoparticle Research, 2013; 15(6): 1676.
66. Cunningham FJ, et al. Nanobiolistics: an emerging genetic transformation approach. Methods in Molecular Biology (Clifton NJ), 2020; 2124: 141–159.
67. Berger M. Nanotechnology strategies for plant genetic engineering. Advanced Materials, 2022; 34(7): e2106945.
68. Demirer G, et al. Nanotechnology to advance CRISPR–Cas genetic engineering of plants. Nature Nanotechnology, 2021; 16: 1–8.

6 Nanomaterial Impact on Genetic Transformation
An Outline

Bipratip Dutta, Sougata Bhattacharjee,
Tilak Mondal, and Rakesh Bhowmick

CONTENTS

DOI: 10.1201/9781003322122-6

6.1 INTRODUCTION

Changing climatic conditions like increasing average temperatures, shrinking fresh-water resources, extreme weather, soil degradation, crop disease infestation, resulting in reduced crop yields and increased food demand for a growing population, present a profound challenge for agricultural productivity and food security in the 21st century [1–3]. By 2050, global food demand will have increased by 60%, along with an estimated population of 9.6 billion [4]. Current crop improvement efforts are mainly focused on sustainably increasing crop yield without excessive use of chemical pesticides and fertilizers and reducing crop disease infestation. Conventional plant-breeding strategies have been successfully used in the past for crop improvement, but these strategies are laborious, time consuming, and challenging [5, 6]. Moreover, conventional plant breeding prohibits the introduction of new traits into the plant that do not currently exist within the species [7, 8]. Therefore, novel plant-breeding technologies need to be utilized to tackle the drawbacks of conventional plant-breeding methods [9, 10]. On this front, plant genetic engineering is a powerful tool that complements traditional plant-breeding methods [11].

Genetic engineering can be used to produce crops with the desired traits to meet the food requirements of the ever-growing population. In the last decade, there have been significant developments in the field of genetic engineering, like the advent of third-generation genome-sequencing techniques, genome editing, and bioengineering [12– 14]. These techniques have been successfully used to develop improved germplasm, crops with higher grain yield, quality, and resistance against biotic and abiotic stresses, increased tolerance to herbicides, and improved nutritional quality as well as climate change [15–18].

Plant genetic engineering techniques have revolutionized biological sciences via precise modifications in the genome of the plants. They involve the manipulation of genetic material or the delivery of new genes of interest into the plant genome.

Strategies to manipulate the genetic material can be categorized into three generations: first-generation strategies include meganuclease and zinc finger nucleases (ZFNs); second-generation strategies include transcription activator-like effector nucleases (TALENs); whereas the third-generation tool includes clustered regularly interspaced short palindromic repeat (CRISPR)/CRISPR-associated protein 9, Cas9) nuclease system [19].

The most popular methods to deliver nucleic acids into the plant include the use of *Agrobacterium* species, particle bombardment, electroporation, microinjection, etc. *Agrobacterium*-mediated gene delivery is the most frequently used method for genetic engineering in plants. Nevertheless, *Agrobacterium* primarily infects dicot plants, although monocot plants have also been engineered through *Agrobacterium* using supervirulent strains and superbinary vectors. Still, in most cases the efficiency of *Agrobacterium* infection is very low [20]. On the other hand, the particle bombardment-based gene delivery method generally damages the target tissue, results in multiple integrations, and requires sophisticated tools, leading to a high cost of operation and limiting its widespread use [21]. Similarly, other nucleic acid delivery methods also have several limitations, like low efficiency, limited cargo types, narrow species range, tissue damages, and are not very successful.

Despite several years of technological advancements in genetic engineering, several plant/crop species of commercial interest either could not be transformed or had extremely low transformation efficiency [22]. The major challenge of nucleic acid delivery into plant cells is the presence of a multi-layered and rigid cell wall made up of cellulosic microfibrils. Therefore, protoplasts (plant cells without a cell wall) have also been explored for gene delivery. However, DNA delivery in protoplast lacks optimized protocols for their maintenance and regeneration of plantlets from protoplast culture, which are not available [23]. Hence, delivering biomolecules across plant cell walls is a major challenge for successful and efficient genetic engineering in plants [24, 25]. In recent years, significant improvements have been made in the field of nanotechnology. Nanotechnology-based nucleic acid delivery in living cells has created immense opportunities for the delivery of biomolecules in plant cells [26–28]. Improvements in nanotechnology have created opportunities to overcome the limitations of conventional genetic engineering methods.

Nanoparticles (NPs) can be made from various materials, including organic and inorganic materials, with a dimension of less than 100 nm [29]. NPs can be categorized into several classes based on their physical and chemical properties, shapes, sizes, and source of materials. Mostly, there are three broad classes of NPs: (1) organic NPs, like liposomes, micelles, dendrimers, or compact polymers; (2) inorganic NPs, like quantum dots, metal (gold, silver, etc.) NPs; and (3) carbon-based NPs, like carbon nanotubes (CNTs), fullerenes, and carbon nanofibers. These NPs can be used in diverse arenas because of their capability to alter their physical, chemical, and biological properties according to the need. NPs have been extensively used in animal cells for nucleic acid or drug delivery [30, 31] as well as in plant cells. Several studies have reported their use in improving agronomic traits as sensing materials [32–34], nano-fertilizers [35, 36], pesticides, herbicides, carriers for controlled release of agrochemicals [37], and as nutrients for crop improvement [28, 38, 39]. NPs can also improve plant adaptation to adverse climatic conditions like different biotic and abiotic stresses [40, 41].

NPs have also been shown to play a role in plant tissue cultures such as callus induction, somatic embryogenesis, organogenesis, and the production of secondary metabolites [32, 42]. These studies clearly show that plant cells can take up the NPs through rigid cell walls, which was the major challenge for successful genetic engineering in plants. However, the mechanism of how NPs enter through a plant cell wall is not yet well understood.

Various NPs, such as silica NPs, CNTs, metal, polymeric, and magnetic NPs (MNPs), have been studied for their potential role in plant genetic engineering. Gene transformation and nucleic acid delivery using nanotechnology are superior to traditional biomolecular approaches mainly because it enhances the transformation efficiency for both transient and stable genetic modifications in various plant species [43–45]. NPs are widely used in nanobiology and gene therapy due to their small size, large surface area, biocompatibility, biodegradability, low toxicity, and low-immunogenic properties [46]. Recently, nanomaterial-mediated gene-delivery system has been developed. High transformation efficiency is achieved without external physical or chemical means in plant cells, showing excellent applications in plant genetic engineering [47]. NPs have begun to facilitate and enhance genetic engineering via an efficient and targeted delivery of plasmids, RNA, and ribonucleoproteins (RNPs). Previous work has shown that some NP formulations undergo passive internalization in plants with DNA, RNA, or protein cargo [48, 49]. NPs as gene delivery carriers were developed and have already been applied in tobacco, corn, *Arabidopsis thaliana*, onion, etc. [47]. The NP-based delivery of cargo in plant cells is an emerging field and is expected to play a considerable role in genetic engineering and genome editing in plant cells.

6.2   EXISTING NANOMATERIALS AVAILABLE FOR TRANSFORMATION

Nanotechnology-based genetic material delivery methods are inexpensive, easy, and robust techniques to transfer genes or other macromolecules into plants with high efficiency and low toxicity [50]. Compared to traditional plant transformation methods, NP-based strategies have several advantages, such as low cytotoxicity, easy operation (e.g., removal of cell wall is unnecessary), species independence, and the ability to deliver a diverse range of biomolecules (e.g., nucleic acids, imaging agents, and regulatory-active molecules). Nanotechnology-based transformation methods have been used to deliver biomolecules and chemicals into cells in both plant and mammalian cell systems [51–54]. However, compared to the mammalian cell system, NP-mediated genetic material delivery is more challenging in the plant system because of the presence of the rigid cell wall [55]. It has been proven that NPs can cross the cell wall barrier and deliver the cargo into the plant cell. Moreover, the NP-mediated plant transformation method is more efficient because NPs protect the genetic cargo from cellular enzymatic degradation [56, 57]. Table 6.1 lists various NP-mediated genetic transformation methods in plants.

Most of these particles need to be less than 20 nm in at least one dimension, e.g., quantum dots [58], nanoclusters [59], carbon nanomaterials [60, 61], and nanowires [62]. In addition, smaller nanomaterials have unique advantages which can achieve suborganelle localization, e.g., chloroplast [60], mitochondrion, nucleus [63]. NPs for gene delivery can be classified according to the base material used, such as carbon-based NPs, silicon-based NPs, metallic NPs, and polymer-based NPs, which can deliver different genetic cargos. For example, CNTs can deliver both RNA and DNA [64, 65], but metallic NPs can only deliver DNA [66], whereas silicon-based NPs can carry DNA and proteins, and polymeric NPs (e.g., polyethylene glycol (PEG) and polyethyleneimine (PEI)) can transfer encapsulated RNA, DNA, and proteins into cells [67–70].

In the case of plant cell transformation, cationic NPs are preferred because they can bind to the negatively charged plant cell wall and perform gene transfer [71]. Some NPs require additional physical methods (e.g., magnetoinfection and electroporation) for gene delivery into plant cells, whereas some NPs, such as silicon carbide whiskers (SCW) and mesoporous silica NPs (MSN), can transfer genetic cargo into the plant without using other physical methods [51]. The SCW-mediated transformation has been successfully used to transform tobacco [72]. However, the

TABLE 6.1
Various Nanoparticle (NP)-Mediated Genetic Engineering Methods in Plants

Nanomaterial type	Genetic material	Plant transformed	Reference
CNTs	• DNA with FITC-labelled SWCNTs	Nicotiana tabacum L. cv. Bright yellow (BY-2) cell	[25]
	• YFP plasmid	Nicotiana tabacum protoplast and leaf	[48]
	• GFP and Cy-3 DNA	Nicotiana benthamiana, Eruca sativa (arugula) leaves and protoplasts	[63]
Gold NP capped MSNs	GFP plasmid	Nicotiana tabacum cotyledons	[142]
Functionalized MSNs	mCherry plasmid	Arabidopsis roots	[51]
Gold plated MSNs	Protein and pDNA	Intact tobacco leaves and maize callus	[143]
Zn-S (metal NP)	Plasmid DNA	Tobacco	[146]
Magnetic NPs	Selectable marker gene plasmid	Cotton pollen	[66]
LDH clay nanosheets	dsRNA	Arabidopsis and tobacco BY-2 cells	[147]
DNA nanostructures	siRNA	Tobacco leaves	[14]
Liposome	CAT gene and tomato yellow leaf curl virus gene	Tobacco and tomato cells	[150]

Note: CNTs, carbon nanotubes; FITC, fluorescein isothiocyanate; SWCNTs, single-walled carbon nanotubes; MSNs, mesoporous silica NPs; LDH, layered double hydroxides; CAT, catalase.

SCW method has one disadvantage: a robust protocol is required for plant regeneration from cell cultures. In a study, polymer NPs were used to introduce siRNA into tobacco protoplasts, suggesting that polymer NPs can also deliver nucleic acids into plant cells and provide an alternative gene knockout mechanism in plant cells [67]. Several NPs can penetrate the cell wall (e.g., CNTs and mesoporous silica), whereas other NPs require chemical or physical pre-treatments, such as gold NPs and MNPs, for genetic cargo delivery into the cells. NPs and other new materials might serve as valuable vehicles for editing systems [73, 74]. Working within this context, Hamada et al. (2017) and Imai et al. (2020) proposed a method involving plant bombardment in which the shoot apical meristems of wheat were used as the target tissue [75, 76]. In this study, gold particles coated with the green fluorescent protein gene construct were delivered into the L2 cell layer of the shoot apical meristems of wheat. This approach provided a stable transformation in wheat without embryogenic callus culture and can be applied to other crops that have not been successfully transformed via conventional methods.

6.2.1 CARBON NANOPARTICLES

Engineered carbon nanomaterials have many excellent applications due to their outstanding mechanical, electrical, optical, and thermal properties. CNTs, carbon dots (CDs), graphene, graphene oxide, and nanodiamonds are the primary materials in the carbon nanomaterial family. So far, most scientific research is focused on the interactions between carbon nanomaterials and mammalian cells [76]. Research on whether and why carbon nanomaterials can be exploited as carriers to deliver foreign genes into plant cells is still ongoing [50, 77].

6.2.1.1 Carbon Dots

CDs were discovered in 2004 [79], and since then, have been extensively used in diverse fields like biomedicine [79, 80], photocatalysis [81], solar energy conversion [82], CD-based light-emitting diodes, supercapacitors, etc. CDs are less than 10 nm in diameter and have no dimension. They also have photoluminescence properties that enhance their cellular uptake, translocation, and accumulation in plants [83]. CDs have an electronegative oxygen species (i.e., hydroxyl or carboxyl) on their surface, making them negatively charged. However, their surface charge can be altered, making them positively charged, which can be electrostatically bound to pDNA and deliver genetic material into a plant cell of several species, like rice, wheat, mung bean. Water-soluble CDs have also been successfully synthesized for DNA absorption and a low-pressure spray method was used to silence endogenous plant genes [84].

6.2.1.2 Carbon Nanotubes

CNTs are long and thin cylindrical molecules composed of sheets of graphene (single-layer carbon atoms) [85]. Single-walled CNTs (SWCNTs) are made of one layer of a graphene sheet and are generally <2 nm in diameter, whereas multi-walled CNTs (MWCNTs) range from 5 to 100 nm in diameter and have more than one layer of graphene sheets. However, both can be several micrometers or even millimeters long [86, 87]. CNTs have high mechanical tensile strength, flexibility, aspect ratio, thermal conductivity, light weight, and other thermal or mechanical properties, which make them ideal candidates for electronic devices, biosensors, transistors, lithium-ion batteries, and various electronic applications [88, 89].

CNTs are generally water-insoluble due to their highly hydrophobic surfaces, but water-soluble CNTs can be synthesized using high-pressure CO conversion [90]. They are positively charged and can attach to nucleic acids through electrostatic interaction. Non-covalent adsorption or wrapping of biomolecules, such as nucleic acids and proteins, occurs via the van der Waals force and π–π stacking [91]. Their reactive surface can be functionalized for loading, and a hollow interior can be filled with various biomolecules to be delivered, such as drugs, peptides, or nucleic acids. The reactive surface of CNTs prevents the loaded cargo from getting degraded or denatured from the surrounding environment [86]. CNTs have extensively been used to deliver foreign DNAs, siRNAs, biomolecules, or other therapeutic agents in mammalian cells [92]. Functionally modified CNTs can effectively deliver genetic cargo at the cellular or subcellular level and exhibit controlled degradation and non-toxicity to the plant cells [93].

6.2.2 Silicon-Based Nanoparticles

Although the second most abundant element in the earth's crust, silicon is not essential for plant growth and development. However, substantial evidence has shown that silica NPs can promote plant tolerance to abiotic/biotic stresses [94]. MSNs have a unique solid framework, porous structure, large surface area, high thermal stability, and the ability to attach different functional groups to their surface. MSNs can be easily functionalized due to the presence of exposed silanol groups on their surface, which not only provides biocompatibility but also makes them stable tools for various biomedical applications like site-specific targeting, drug loading, etc. [95, 96]. After surface modification, it has been found that MSNs can deliver DNA into mammalian cells. The use of MSNs for biomolecule delivery in the plant cell is also being explored. To deliver biomolecules to plant cells, MSNs have to be surface-coated by different polymers like PEI, poly(allylamine hydrochloride), or polyamidoamine dendrimers. This surface coating generates a positive charge on the surface of the MSNs, helping in the binding of the biomolecules. The mesoporous structure of the MSNs is particularly advantageous, as it allows the loading of a large amount of DNA/other materials on them and the release of the cargo on the target sites can be performed in a controlled manner.

6.2.3 Metal Nanoparticles

Metal NPs, such as gold, silver, platinum, and palladium, possess unique optical absorption and scattering properties arising from localized surface plasmon resonance that makes them very important for bioimaging or biomedical applications [97–99]. Gold and silver NPs are of particular interest because they are flexible with their size and shape, surface modification, bio-conjugation, and optical properties

[100–102]. Metal NPs have been broadly used for intracellular delivery of diverse molecules like DNA, drugs, peptides, etc., in animal/mammalian cells [103]. Though in plant cells, biomolecule delivery through metal NPs is still limited because of the presence of the cell wall. However, meaningful progress has been made to use metal NPs in plant science as nano-fertilizers or micronutrients for crop improvements [34, 35], Mostly, metal NPs have been used as capping agents with other nanomaterials and as a trigger for cargo release.

6.2.4 MAGNETIC NANOPARTICLES

One very effective way of nucleic acid delivery is by magnetofication, i.e., nucleic acid delivery guided by the magnetic field. The metal NPs used in this process also possess magnetic properties. MNPs generally include metal NPs (nickel or cobalt), metal oxides (iron oxides), and metal alloys (FeCo, FePt, etc.). MNPs have excellent biocompatibility and superparamagnetism, i.e., they show magnetism only in the presence of a magnetic field [104, 105]. Positively charged MNPs generally deliver nucleic acid into animal and plant cells [52, 106] in a non-assisted manner. Superparamagnetic properties of NPs shield the cargo from digestion, enhance nucleic acid loading capacity, give better penetration, have a significant cost reduction, and enable the delivery of biomolecules when exposed to an external magnetic field [107, 108]. However, only a few articles have been published on magnetofection with plant cells, and this technology is still controversial.

6.2.5 LAYERED DOUBLE HYDROXIDE

Layered double hydroxide (LDH) is a class of ionic lamellar inorganic materials with a positive charge on their surfaces; they occur in nature as minerals. They have been explored in the field of drug delivery [109], water oxidation [110], catalysis [111], supercapacitors [112], gas absorbents [113], and analytical extractions [114]. LDHs have a positive charge on their surface and thus can bind to nucleic acids by electrostatic interaction. LDH can deliver DNA into a plant cell by three possible methods:

1. The plant cell wall prevents DNA/RNA–LDH hybrids from reaching the plasma membrane and cytoplasm, but the bound DNA/RNA can be dissociated at the cell wall and reach the plasma membrane cytoplasm.
2. The DNA/RNA–LDH hybrid can pass through the plasma membrane by a non-endocytic pathway.
3. The DNA/RNA–LDH hybrid can be internalized into the cell by endocytosis.

LDH has also been used to deliver dsRNA and pDNA into plant cells, making them tolerant against different biotic stresses.

6.2.6 DNA NANOSTRUCTURES

DNA nanostructure comprises the two most interesting emerging fields: DNA (gene) technology and nanotechnology. Being made up of nucleic acids, DNA nanostructures have many advantages, like nanoscale size, the programmable and modular nature of DNA base pairing, which can create a variety of custom predesigned shapes from the simplest immobile Holliday junctions to DNA origami, two-dimensional shapes, and complex three-dimensional shapes. DNA nanostructures of different sizes and shapes have been synthesized and shown significant importance for delivering drugs, DNA, RNA, and proteins in animal systems [115, 116]. However, DNA nanostructures have rarely been explored in plant systems. Zhang et al. (2020) used DNA nanostructure to deliver siRNA into plants for transgene-free gene-silencing applications [117]. DNA nanostructures have high biocompatibility, and their delivery of nucleic acid into plant or animal cells depends largely on the size and shape of the nanostructure and the binding nature of the nucleic acid with it.

6.2.7   Liposomes

Liposomes are vesicle-like structures composed of a phospholipid bilayer surrounding an inner aqueous core [117]. Liposome molecules can be positively or negatively charged, or may even be uncharged. Until now, liposome-mediated nucleic acid delivery has been most suitable for protoplast, and the transformation efficiency has been observed to be very high in the presence of PEG and Ca^{2+} [118]. Three pathways show liposome-mediated genetic transformation in plants:

1. Liposome-protoplast fusion model, especially in the presence of PEG or poly-vinyl alcohol in combination with Ca^{2+}
2. Chemical endocytosis is the endocytosis of liposome-encapsulated biomolecule by the protoplast, although this uptake is more effective under the influence of fusogens [119]
3. Leakage of lysosomal contents after coming into contact with the plasma membrane. This model allows the nucleic acid to pass through the membrane by forming transient pores. In addition to protoplast, liposomes have been found to deliver biomolecules into pollen grains [120]. To date, there is no example of liposome-mediated delivery across the cell wall.

6.2.8   Other Nanomaterials

6.2.8.1   Polymer-Based Nanoparticles

Several polymer-based NPs can be used to deliver biomolecules into the cells. Some of them are discussed here:

- Positively charged conjugated polymer NPs (CPNs) are biocompatible, with flexible structure and function [121]. The conjugated backbone provides optical properties, whereas the positive charge helps to bind with nucleic acids. This complex can also protect the biomolecule from degradation. CPNs have been used to deliver siRNA vectors to achieve effective gene silencing in protoplast.
- Chitosan is a deacetylated derivative of chitin, mainly extracted from shellfish. It is widely used in biomedical engineering, agriculture, food, and other fields [122, 123]. Chitosan has been used extensively for gene transfer in mammalian

[124, 125] and animal cells [126]. However, in plant cells, its use in the genetic material delivery mechanism has not yet been explored, although it has been used to deliver plant nutrients [127]. Being positively charged, it can bind with negatively charged nucleic acids and protect them from degradation; therefore, it has potential as a plant gene delivery carrier.

- Dendritic polymers or dendrimers are highly branched polymers with a positive charge. Dendrimers are positively charged and have quaternary ammonium and guanidine groups which help to load the nucleic acids.
- Starch is extracted from agricultural raw materials, and is non-toxic, biocompatible, and biodegradable. It can adsorb nucleic acid through electrostatic interaction [128]. Extra assistance is needed to deliver the nucleic acids inside the plant cells, as they are large. Although DNA-loaded starch particles are highly stable, the efficiency of successful transformation is very low. Till now, starch particles have rarely been used in plant cells.

6.2.8.2   Peptide-Based Nanoparticles

Peptide-based NPs are non-viral carriers that are generally polycationic and bind to charged DNA and negatively protect them from enzymatic degradation [129]. Cell-penetrating peptides are primarily studied in animal cells. These are short peptides that can carry small chemical compounds, large DNA molecules, and nano-sized particles [130]. But peptide NPs can be easily captured and degraded along with the biomolecule in the endosomes, resulting in low efficiency. So, additional strategies are needed to escape from that endosomal trap, and it is assumed that peptides containing histidine-rich amino acids can escape endosomal degradation in mammalian cells [131]. Peptide NP-based gene delivery mainly depends on the number of cell-penetrating peptides and is less affected by physiological factors [132].

6.3   EXAMPLES OF NANOMATERIAL-MEDIATED TRANSFORMATION

6.3.1   CARBON NANOPARTICLES

Carbon nanomaterials are one of the most exploited NPs in both animal and plant genetic engineering. They have exceptional optical, electrical, mechanical, and thermal properties, which render them very useful in genetic engineering and other fields. Carbon NPs include CDs, CNTs, graphene, nanodiamond, and fullerene. Among these, CNTs are the most used in genetic engineering in both animal and plant cells.

6.3.1.1   Carbon Dots

The fluorescent properties of CDs have made them amenable for enhanced cellular uptake, translocation, and accumulation in plants. Besides genetic cargo delivery, CDs have been observed to affect plant growth, development, photosynthesis, and resistance to abiotic/biotic stresses [133]. The negative charge on the surface of CDs was altered to a positive charge by introducing PEI to the surface, and this positively charged CD can electrostatically bind to pDNA, resulting in efficient genetic transformation in several species, like rice, wheat, and mung bean. Schwartz et al. used PEI as the carbon source, successfully synthesized water-soluble CDs for DNA adsorption, and developed a low-pressure spray method to silence endogenous plant genes encoding magnesium chelatase [85].

6.3.1.2 Carbon Nanotubes

CNTs are among the most used NPs in diverse fields like electronics, solar cells, bioengineering, biosensors, drug delivery, and gene delivery. [76]. The most preferred method of CNT production is chemical vapor deposition [134]. Protein, carbohydrate, and nucleic acid (DNA, RNA) functionalization with CNT improves water dispersibility, biocompatibility, and low toxicity, thus providing a new platform for green nanotechnology [135]. Among the biomolecules, DNA has a well-defined length and sequence and high dispersion efficiency, and is covalently or non-covalently functionalized with CNTs [136]. Nucleic acids bind wrap on CNT by van der Waals and π–π stacking interactions.

Liu et al. found that oxidized SWCNTs can penetrate through the plant cell wall in *Nicotiana tabacum* L. cv. Bright Yellow (BY-2) cell [25]. They found that DNA and fluorescein isothiocyanate (FITC)-labeled SWCNTs successfully passed through the intact plant cell without an external aid (e.g., particle bombardment). When those tobacco cells were treated with an inhibitor of endocytosis, the FITC signal significantly decreased, suggesting endocytosis as a possible mechanism of SWCNT entry into the plant cell.

The PEI-modified SWCNTs could adsorb nucleic acids through electrostatic interaction and protect them from nuclease digestion [49]. They also showed that DNA associated with PEI SWCNTs could be delivered into mature plant cells.

Kwak et al. (2019) designed chitosan-complexed SWCNTs, which successfully delivered pDNA to *Arabidopsis* mesophyll chloroplasts without external aid [60]. Demirer et al. constructed a siRNA-SWCNT delivery system using π–π stacking [138], which can protect the RNA from degradation and confer endogenous *GFP* gene silencing when delivered into plant cells. Burlaka et al. optimized the concentration of non-covalently functionalized SWCNTs and MWCNTs and found that SWCNTs can successfully cross plant cell walls whereas MWCNTs cannot [48]. Wong et al. (2016) developed a mathematical model, lipid exchange envelope and penetration (LEEP), while studying the impact of NP size and zeta potential on penetrating plant membrane [139]. The model predicted that the size of the NP is essential for penetration through the membrane, and a size below the critical limit will fail to penetrate the membrane even at different zeta potential.

6.3.2 Silicon-Based Nanoparticles

MSNs provide good biocompatibility, which helps in drug loading and site-specific targeting. MSNs have been used to deliver nucleic acids in animal cell lines/tissues [140, 141]. MSNs can also be internalized in plant cells. Hussain et al. (2013) found that amine cross-linked FITC-functionalized MSNs could be effectively absorbed by wheat, lupin, and *Arabidopsis* [142]. Torney et al. had designed honeycomb MSNs with a diameter of 3 nm to transport nucleic acids and chemicals into plant protoplast and intact leaves [143]. They also designed MSNs with gold-capped NPs (Au-MSNs), which efficiently delivered pDNA and was expressed in intact tobacco leaves and maize callus. Martin-Ortigosa et al. (2012) co-delivered protein and pDNA into plant cells with Au-MSNs via the biolistic method [144]. Chang et al. (2013) developed a gene delivery system using organic group functionalized MSNs, which delivered DNA into intact *Arabidopsis* roots without any external force [51]. Hajiahmadi et al. (2019) reported that DNA could be loaded on to MSNs and sprayed or injected into tomato leaves/shoots [145]. Plasmids having *GUS* gene or *Cry1Ab* gene were loaded on to the MSNs, and reverse transcriptase polymerase chain reaction and Western blot assay later confirmed their expression. These findings revealed that MSNs could deliver DNA molecules into plant cells without the help of any external force.

6.3.3 Metal Nanoparticles

Metal NPs have been used in plant science as nano-fertilizers or micronutrients, but their use to deliver biomolecules in plant cells is minimal. Vijayakumar et al. (2010) used carbon bullets functionalized with gold NPs to deliver biomolecules [146]. These particles, embedded in the carbon matrix, can carry up to 2400 ng of plasmid DNA. Fu and colleagues use Zn-S NPs to deliver DNA into plant cells [147]. Plasmid-coated Zn-S NPs were delivered by ultrasound, producing transgenic tobacco plants. However, metal NPs are mostly used for bioimaging and trigger cargo release because of their poor loading capacity.

6.3.4 Magnetic Nanoparticles

MNPs are generally metal NPs that have magnetic properties. The delivery of nucleic acid through MNPs is called magnetofication. Hao and his co-workers showed the use of magnetic gold NPs (mGNPs) in genetic material delivery [107]. mGNPs were surface-functionalized, and FITC-labeled, plasmid DNA was bound to the surface by an amide linkage. Then the external magnetic field was applied to deliver the DNA into canola protoplasts. Up to 95% of efficiency in pDNA delivery was reported. Zhao et al. reported the MNP-based DNA delivery method "pollen magnetofication," in which PEI-functionalized MNPs were used to deliver DNA into cotton pollen [67]. Magnetofected pollen was then used to pollinate plants, resulting in the production of transgenic plants without any tissue culture step.

6.3.5 Layered Double Hydroxide

LDH is made of inorganic materials that have been used in genetic material delivery in plants. Efficient nucleic acid delivery has been achieved in the cytoplasm and nucleus of *Arabidopsis* and tobacco BY-2 cells [148]. Mitter et al. (2017) synthesized LDH clay nanosheets with dsRNA, which can be retained in the leaves for a long time. Further, it was observed that LDH-dsRNA provided more excellent protection against viral infections. Liu et al. successfully created pDNA bound with non-toxic clay nanosheets, which, when sprayed on to the plant leaves, could release pDNA stably for more than 35 days, inhibiting tomato yellow leaf curl virus [25].

6.3.6 DNA NANOSTRUCTURES

Exploring DNA nanostructure in genetic material delivery in plant cells is very limited. Zhang et al. (2019) used DNA nanostructure for siRNA delivery in plants [14]. These nanostructures were loaded with siRNA targeting a *GFP* gene and delivered into tobacco leaves for a gene-silencing experiment. The researchers assumed that these gene-silencing mechanisms were different from the usual, and later found that different nanostructures activated different silencing pathways.

6.3.7 LIPOSOMES

Liposome-mediated DNA delivery is only successful in the protoplast in the plant cell. Karny and colleagues (2018) found that liposomes based on HSPC (L-α-phosphatidylcholine) can penetrate the leaves of plants after foliar spraying and subsequently release the biomolecule [149]. Nagata et al. (1981) studied the interaction between plant protoplast and liposomes [150]. They used differently charged liposomes and found that liposomes are integrated into the plant cell through endocytosis. Ahokas demonstrated that positively charged liposomes could deliver DNA into pea pollen grains [121]. Rosenberg and co-workers used negatively charged liposomes to deliver the *CAT* (chloramphenicol acetyltransferase) gene and tomato yellow leaf curl virus gene into tobacco and tomato cells [151]. However, there is no example of liposome-mediated genetic transformation across intact cell walls of plants.

Besides these, different polymer- and peptide-based nanomaterials were used to transfer genetic material into both plant and animal cells.

6.4 COMPARISON OF NANOMATERIAL-MEDIATED GENETIC TRANSFORMATION WITH CONVENTIONAL GENETIC TRANSFORMATION

The genetic engineering technique modifies plant genomes, including inserting new genes either from a foreign source or from the same source, stopping the expression of an existing gene by knockout or knockdown approach or genome editing. These modifications depend on the efficient delivery of the biomolecules (DNA/RNA/RNPS/small RNAs, etc.) to the targeted plant cells [151, 152]. Available genetic transformation methods, listed in Table 6.2, have several limitations, which decrease the efficiency of genetic transformation, such as damage to the plant tissues, non-specific integration, non-significant gene expression, tissue specificity, species specificity, etc. [12]. Most of these techniques can be used in a minimal host range, and they cause post-modification regeneration and fertility problems in transgenic plants. Methods like *Agrobacterium*-mediated transformation and gene gun-based biolistic approaches can lead to important host plant gene silencing, making them unusable. Although these techniques have evolved a lot and have had numerous successes,

TABLE 6.2
Traditional Genetic Engineering Methods in Plants

Methods	Characteristics	Limitations	Reference
Agrobacterium- **mediated transformation**	*Agrobacterium* Ti plasmid naturally infects plant cells and delivers the gene of interest into the nucleus of the host genome	• Narrow host range, mostly dicots • Poor transformation efficiency • Regeneration protocols required	[164] [165]
Viral vector- mediated transformation	Virus-based vectors are generally used for transient expression	• Highly species-specific • DNA integration is not stable • Limited cargo capacity	[166] [24]
Biolistics (gene gun)	DNA-coated gold or tungsten particles are bombarded over the tissues, cells, or plant organs	• Due to its high velocity it damages the tissue • Multiple integration of transgene • Regeneration protocol required • Random integration	[167] [168]
Electroporation	Electrical current applied through pollen grains or protoplasts	• Host-specific • Damages the target tissue • Chances of deletion	[169] [170]
Microinjection	DNA is directly injected into the cell using a needle or micropipette	• Appropriate for large cells • Low efficiency	[171]

Agrobacterium-mediated transformation can now be used for monocot plants. Genetic engineering has progressed immensely with the development of gene-editing techniques like ZFN, TALEN, and especially CRISPR/Cas.

But still, there are certain complications that limit genetic-engineering techniques. One of the main complications is related to plant cell structure. Compared to animal cells, plant cells have an extra layer outside the plasma membrane, a rigid cell wall, which is impervious to most of the ions and molecules. The cell wall acts as a physical barrier from the outer environment and maintains the rigidity and shape of the cell [153]. A plant cell wall is mainly composed of cellulose, hemicellulose, lignin, etc., and has a pore size varying from 3.5 to 5.2 nm [154]. Because of this small pore size, most genetic material cannot pass through the cell wall barrier.

On the other hand, NP-based genetic-engineering techniques have emerged as one of the best genetic material delivery systems because of their colossal diversity, ease of use, and success [13, 24]. NP-mediated genetic material delivery in plant cells has huge potential and has revolutionized genetic-engineering techniques [155,156]. In these methods, the biomolecule to be delivered is attached to the NPs either covalently or non-covalently and transferred to the plant cells without causing damage to the host tissue. Due to their small size, NPs can pass through the cell wall and cell membrane to deliver the biomolecule inside the plant cell. NPs can enter through the cell walls, diffuse through the plasmodesmata, and migrate through the xylem. Some studies have also suggested their movement through apoplast and symplast.

Although there are certain drawbacks which hinder the use of NPs in genetic-engineering technologies, one of them is nanophytotoxicity [157]. Nanophytotoxicity is the negative effect of NPs on plant growth ad development, causing damage to the plant or the environment. Studies have shown that uptake of NPs can block the plant vascular system, resulting in structural damage to plants and their genetic material [158, 159]. The plants' reproductive growth has been negatively affected by silver NPs [160]. Inside the plant cells, NP deposition can result in increased reactivity and instability [37]. NPs that have high oxidative properties can disturb normal cell metabolism and interfere with genetic regulation [159, 161].

Another challenge for NP-mediated genetic engineering is the efficient binding of the biomolecules to the NPs and releasing that biomolecule inside plant cells [162, 163]. The binding of the biomolecule with the NPs largely depends on the size, shape, surface charge, structure, chemical composition, etc. NPs have to be modified and designed to deliver the definite cargo to the designated cell organelle, which is one of the limitations of traditional genetic-engineering techniques. NPs can also be used across tissues and species, they are mostly biocompatible, and most do not need any special equipment.

Despite being in its infancy, NP-mediated genetic material delivery has tremendous potential, and with ongoing research, this technology will be further developed.

6.5 CONCLUSIONS

With the increasing demand for food, plant genetic engineering is the fastest way to generate improved crops with better nutrition, increased yield, disease resistance, and stress resistance. Therefore, establishing efficient plant genetic transformation methods is an essential area of research in plant science. In this regard, nanotechnology will definitely play a critical role in the design of smart crop systems. Nanotechnology has been used in nutrient supply, providing protection against biotic and abiotic stresses, and increasing the amount of enzymatic and non-enzymatic compounds and genetic supplies. NPs have been used as nanograms, nano-pesticides, nano-fertilizers, nanosensors, nano-chips, and many more. NP-mediated plant genetic transformation has several advantages over traditional genetic transformation methods. NPs can pass through the plant cell wall and the organellar membrane, and can transfer large segments of genetic material into the plant cell while protecting them from degradation. These NPs have no host restriction and are biocompatible; they can be tuned according to the researcher's need. So, this NP-mediated genetic material delivery will shortly be essential in producing innovative crop systems. However, more research has to be undertaken, and all the issues regarding NP-mediated genetic material delivery should be resolved, such as how the NPs internalized into plant cells are still largely uncharacterized; for better efficiency, this needs to be addressed.

In some instances, NP internalization in plant cells is not necessary; how exactly this process occurs is still unknown. The effect of NP-mediated phytotoxicity has to be resolved soon to produce genetically improved crops without any risk of toxicity. Recent genome-editing techniques like CRISPR/Cas have been explored for the production of genetically modified crops. Whether these genome-editing techniques can be improved with nanotechnology or not is still an area for considerable research. Furthermore, previously developed technologies like speed-breeding and speed-editing strategies could be developed to speed up the genetic-engineering process by incorporating NPs to establish crops with desired traits.

In summary, nanotechnology for plant genetic transformation has enormous potential, and in the near future, this may be the driving force for generating better crops. But before that, more study is needed in this field to resolve the issues and to come up with an efficient, robust strategy to create crops with desired traits.

REFERENCES

1. Rosenzweig C, Elliott J, Deryng D, Ruane AC, Müller C, Arneth A, Boote KJ, Folberth C, Glotter M, Khabarov N, Neumann K. Assessing agricultural risks of climate change in the 21st century in a global gridded crop model intercomparison. Proceedings of the National Academy of Sciences, 2014; 111(9): 3268–3273.
2. Vos R, Bellù LG. Global trends and challenges to food and agriculture into the 21st century. Sustainable Food and Agriculture, 2019; 1: 11–30.
3. Shaheen A, Abed Y. Knowledge, attitude, and practice among farmworkers applying pesticides in cultivated area of the Jericho district: a cross-sectional study. The Lancet, 2018; 391(1): S3. doi.org/10.1016/S0140-6736(18)30328-3
4. Bajželj B, Richards KS, Allwood JM, Smith P, Dennis JS, Curmi E, Gilligan CA. Importance of food-demand management for climate mitigation. Nature Climate Change, 2014; 4(10): 924–929.
5. Prohens J. Plant breeding: a success story to be continued thanks to the advances in genomics. Frontiers in Plant Science, 2011; 2(51): 1–3. doi.org/10.3389/fpls.2011.00051
6. Borlaug NE. Contributions of conventional plant breeding to food production. Science, 1983; 219(4585): 689–693.
7. Bhargava A, Srivastava S. Participatory Plant Breeding: Concept and Applications. Springer, Singapore. 2019.
8. Morris ML, Bellon MR. Participatory plant breeding research: opportunities and challenges for the international crop improvement system. Euphytica, 2004; 136(1): 21–35.
9. Chen K, Wang Y, Zhang R, Zhang H, Gao C. CRISPR/Cas genome editing and precision plant breeding in agriculture. Annual Review of Plant Biology, 2019; 70(1): 667–697.
10. Fiaz S, Khan SA, Anis GB, Gaballah MM, Riaz A. CRISPR/Cas techniques: a new method for RNA interference in cereals. CRISPR and RNAi Systems, 2021; 233–252.
11. Datta A. Genetic engineering for improving quality and productivity of crops. Agriculture and Food Security, 2013; 2(1): 1–3.
12. Altpeter F, Springer NM, Bartley LE, Blechl AE, Brutnell TP, Citovsky V, Conrad LJ, Gelvin SB, Jackson DP, Kausch AP, Lemaux PG. Advancing crop transformation in the era of genome editing. The Plant Cell, 2016; 28(7): 1510–1520.

13. Wang JW, Grandio EG, Newkirk GM, Demirer GS, Butrus S, Giraldo JP, Landry MP. Nanoparticle-mediated genetic engineering of plants. Molecular Plant, 2019; 12(8): 1037–1040.

14. Zhang H, Demirer GS, Zhang H, Ye T, Goh NS, Aditham AJ, Cunningham FJ, Fan C, Landry MP. DNA nanostructures coordinate gene silencing in mature plants. Proceedings of the National Academy of Sciences, 2019; 116(15): 7543–7548.

15. Kissoudis C, van de Wiel C, Visser RG, van der Linden G. Enhancing crop resilience to combined abiotic and biotic stress through the dissection of physiological and molecular crosstalk. Frontiers in Plant Science, 2014; 5: 207. doi.org/10.3389/fpls.2014.00207

16. Dong OX, Ronald PC. Genetic engineering for disease resistance in plants: recent progress and future perspectives. Plant Physiology, 2019; 180(1): 26–38.

17. Hilder VA, Boulter D. Genetic engineering of crop plants for insect resistance – a critical review. Crop Protection, 1999; 18(3): 177–191.

18. Bonny S. Genetically modified herbicide-tolerant crops, weeds, and herbicides: overview and impact. Environmental Management, 2016; 57(1): 31–48.

19. Ahmar S, Mahmood T, Fiaz S, Mora-Poblete F, Shafique MS, Chattha MS, Jung KH. Advantage of nanotechnology-based genome editing system and its application in crop improvement. Frontiers in Plant Science, 2021; 12: 663–849.

20. Gelvin SB. Integration of *Agrobacterium* T-DNA into the plant genome. Annual Review of Genetics, 2017; 51: 195–217.

21. Sanford JC. Biolistic plant transformation. Physiologia Plantarum, 1990; 79(1): 206–209.

22. Kempken F, Jung C, editors. Genetic Modification of Plants: Agriculture, Horticulture and Forestry. Springer Science and Business Media, Berlin. 2009; 15.

23. Eeckhaut T, Lakshmanan PS, Deryckere D, Van Bockstaele E, Van Huylenbroeck J. Progress in plant protoplast research. Planta, 2013; 238(6): 991–1003.

24. Cunningham FJ, Goh NS, Demirer GS, Matos JL, Landry MP. Nanoparticle-mediated delivery towards advancing plant genetic engineering. Trends in Biotechnology, 2018; 36(9): 882–897.

25. Liu Q, Chen B, Wang Q, Shi X, Xiao Z, Lin J, Fang X. Carbon nanotubes as molecular transporters for walled plant cells. Nano Letters, 2009; 9(3): 1007–1010.

26. Riley MK, Vermerris W. Recent advances in nanomaterials for gene delivery – a review. Nanomaterials, 2017; 7(5): 94.

27. Sanzari I, Leone A, Ambrosone A. Nanotechnology in plant science: to make a long story short. Frontiers in Bioengineering and Biotechnology, 2019; 7: 120.

28. Wang P, Lombi E, Zhao FJ, Kopittke PM. Nanotechnology: a new opportunity in plant sciences. Trends in Plant Science, 2016; 21(8): 699–712.

29. Khan I, Saeed K, Khan I. Nanoparticles: properties, applications and toxicities. Arabian Journal of Chemistry, 2019; 12(7): 908–931.

30. Shan Z, Jiang Y, Guo M, Bennett JC, Li X, Tian H, Oakes K, Zhang X, Zhou Y, Huang Q, Chen H. Promoting DNA loading on magnetic nanoparticles using a DNA condensation strategy. Colloids and Surfaces B: Biointerfaces, 2015; 125: 247–254.

31. Jat SK, Selvaraj D, Muthiah R, Bhattacharjee RR. A self-releasing magnetic nanomaterial for sustained release of doxorubicin and its anticancer cell activity. Chemistry Select, 2018; 3(46): 13123–13131.

32. Verma ML, Kumar P, Sharma D, Verma AD, Jana AK. Advances in nanobiotechnology with special reference to plant systems. *In*: Plant Nanobionics. Springer, Cham. 2019; 371–387.

33. Chaudhry N, Dwivedi S, Chaudhry V, Singh A, Saquib Q, Azam A, Musarrat J. Bio-inspired nanomaterials in agriculture and food: current status, foreseen applications and challenges. Microbial Pathogenesis, 2018; 123: 196–200.

34. Duhan JS, Kumar R, Kumar N, Kaur P, Nehra K, Duhan S. Nanotechnology: the new perspective in precision agriculture. Biotechnology Reports, 2017; 15: 11–23.

35. DeRosa MC, Monreal C, Schnitzer M, Walsh R, Sultan Y. Nanotechnology in fertilizers. Nature Nanotechnology, 2010; 5: 91. doi.org/10.1038/nnano.2010.2.

36. Liu R, Lal R. Potentials of engineered nanoparticles as fertilizers for increasing agronomic productions. Science of the Total Environment, 2015; 514: 131–139.

37. Lv J, Christie P, Zhang S. Uptake, translocation, and transformation of metal-based nanoparticles in plants: recent advances and methodological challenges. Environmental Science: Nano, 2019; 6(1): 41–59.

38. Medina-Pérez G, Fernández-Luqueño F, Campos-Montiel RG, Sánchez-López KB, Afanador-Barajas LN, Prince L. Nanotechnology in crop protection: status and future trends. Nano-Biopesticides Today and Future Perspectives, 2019; 17–45.

39. Shang Y, Hasan MK, Ahammed GJ, Li M, Yin H, Zhou J. Applications of nanotechnology in plant growth and crop protection: a review. Molecules, 2019; 24(14): 2558.

40. Torabian S, Zahedi M, Khoshgoftar AH. Effects of foliar spray of nano-particles of $FeSO_4$ on the growth and ion content of sunflower under saline condition. Journal of Plant Nutrition, 2017; 40(5): 615–623.

41. Tripathi DK, Singh S, Singh VP, Prasad SM, Dubey NK, Chauhan DK. Silicon nanoparticles more effectively alleviated UV-B stress than silicon in wheat (*Triticum aestivum*) seedlings. Plant Physiology and Biochemistry, 2017; 110: 70–81.

42. Kim DH, Gopal J, Sivanesan I. Nanomaterials in plant tissue culture: the disclosed and undisclosed. RSC Advances, 2017; 7(58): 36492–36505.

43. Serag MF, Kaji N, Habuchi S, Bianco A, Baba Y. Nanobiotechnology meets plant cell biology: carbon nanotubes as organelle targeting nanocarriers. RSC Advances, 2013; 3(15): 4856–4862.

44. Vanhaeren H, Inzé D, Gonzalez N. Plant growth beyond limits. Trends in Plant Science, 2016; 21(2): 102–109.

45. Novák O, Napier R, Ljung K. Zooming in on plant hormone analysis: tissue- and cell-specific approaches. Annual Review of Plant Biology, 2017; 68: 323–348.

46. Yin H, Kanasty RL, Eltoukhy AA, Vegas AJ, Dorkin JR, Anderson DG. Non-viral vectors for gene-based therapy. Nature Reviews Genetics, 2014; 15(8): 541–555.

47. Jat SK, Bhattacharya J, Sharma MK. Nanomaterial based gene delivery: a promising method for plant genome engineering. Journal of Materials Chemistry B, 2020; 8(19): 4165–4175.

48. Burlaka OM, Pirko YV, Yemets AI, Blume YB. Plant genetic transformation using carbon nanotubes for DNA delivery. Cytology and Genetics, 2015; 49(6): 349–357.

49. Demirer GS, Chang R, Zhang H, Chio L, Landry MP. Nanoparticle-guided biomolecule delivery for transgene expression and gene silencing in mature plants. Biophysical Journal, 2018; 114(3): 217a.

50. Chandrasekaran R, Rajiv P, Abd-Elsalam KA. Carbon nanotubes: plant gene delivery and genome editing. Carbon Nanomaterials for Agri-Food and Environmental Applications, 2020; 1: 279–296.

51. Chang FP, Kuang LY, Huang CA, Jane WN, Hung Y, Yue-ie CH, Mou CY. A simple plant gene delivery system using mesoporous silica nanoparticles as carriers. Journal of Materials Chemistry B, 2013; 1(39): 5279–5287.

52. Wang Y, Cui H, Li K, Sun C, Du W, Cui J, Zhao X, Chen W. A magnetic nanoparticle-based multiple-gene delivery system for transfection of porcine kidney cells. PloS One, 2014; 9(7): e102886.

53. Mahakham W, Sarmah AK, Maensiri S, Theerakulpisut P. Nanopriming technology for enhancing germination and starch metabolism of aged rice seeds using phytosynthesized silver nanoparticles. Scientific Reports, 2017; 7(1): 1–21.

54. Fortuni B, Inose T, Ricci M, Fujita Y, Van Zundert I, Masuhara A, Fron E, Mizuno H, Latterini L, Rocha S, Uji-i H. Polymeric engineering of nanoparticles for highly efficient multifunctional drug delivery systems. Scientific Reports, 2019; 9(1): 1–3.

55. Mao Y, Botella JR, Liu Y, Zhu JK. Gene editing in plants: progress and challenges. National Science Review, 2019; 6(3): 421–437.

56. Finiuk N, Buziashvili A, Burlaka O, Zaichenko A, Mitina N, Miagkota O, Lobachevska O, Stoika R, Blume Y, Yemets A. Vestigation of novel oligoelectrolyte polymer carriers for their capacity of DNA delivery into plant cells. Plant Cell, Tissue and Organ Culture (PCTOC), 2017; 131(1): 27–39.

57. Joldersma D, Liu Z. Plant genetics enters the nanoage? Journal of Integrative Plant Biology, 2018; 60(6): 446–447.

58. Pagano L, Maestri E, White JC, Marmiroli N, Marmiroli M. Quantum dots exposure in plants: minimizing the adverse response. Current Opinion in Environmental Science & Health, 2018; 6: 71–76.

59. Zhang H, Cao Y, Xu D, Goh NS, Demirer GS, Cestellos-Blanco S, Chen Y, Landry MP, Yang P. Gold-nanocluster-mediated delivery of siRNA to intact plant cells for efficient gene knockdown. Nano Letters, 2021; 21(13): 5859–5866.

60. Kwak SY, Lew TT, Sweeney CJ, Koman VB, Wong MH, Bohmert-Tatarev K, Snell KD, Seo JS, Chua NH, Strano MS. Chloroplast-selective gene delivery and expression in planta using chitosan-complexed single-walled carbon nanotube carriers. Nature Nanotechnology, 2019; 14(5): 447–455.

61. Li Y, Xu X, Wu Y, Zhuang J, Zhang X, Zhang H, Lei B, Hu C, Liu Y. A review on the effects of carbon dots in plant systems. Materials Chemistry Frontiers, 2020; 4(2): 437–448.

62. Majumdar S, Long RW, Kirkwood JS, Minakova AS, Keller AA. Unraveling metabolic and proteomic features in soybean plants in response to copper hydroxide nanowires compared to a commercial fertilizer. Environmental Science & Technology, 2021; 55(20): 13477–13489.

63. Demirer GS, Zhang H, Matos JL, Goh NS, Cunningham FJ, Sung Y, Chang R, Aditham AJ, Chio L, Cho MJ, Staskawicz B. High aspect ratio nanomaterials enable delivery of functional genetic material without DNA integration in mature plants. Nature Nanotechnology, 2019; 14(5): 456–464.

64. Bates K, Kostarelos K. Carbon nanotubes as vectors for gene therapy: past achievements, present challenges and future goals. Advanced Drug Delivery Reviews, 2013; 65(15): 2023–2033.

65. Beg S, Rizwan M, Sheikh AM, Hasnain MS, Anwer K, Kohli K. Advancement in carbon nanotubes: basics, biomedical applications and toxicity. Journal of Pharmacy and Pharmacology, 2011; 63(2): 141–163.

66. Karimi M, Solati N, Ghasemi A, Estiar MA, Hashemkhani M, Kiani P, Mohamed E, Saeidi A, Taheri M, Avci P, Aref AR. Carbon nanotubes part II: a remarkable carrier for drug and gene delivery. Expert Opinion on Drug Delivery, 2015; 12(7): 1089–10105.

67. Zhao X, Meng Z, Wang Y, Chen W, Sun C, Cui B, Cui J, Yu M, Zeng Z, Guo S, Luo D. Pollen magnetofection for genetic modification with magnetic nanoparticles as gene carriers. Nature Plants, 2017; 3(12): 956–964.

68. Silva AT, Nguyen A, Ye C, Verchot J, Moon JH. Conjugated polymer nanoparticles for effective siRNA delivery to tobacco BY-2 protoplasts. BMC Plant Biology, 2010; 10(1): 1–4.

69. Moon JH, Mendez E, Kim Y, Kaur A. Conjugated polymer nanoparticles for small interfering RNA delivery. Chemical Communications, 2011; 47(29): 8370–8372.

70. Su X, Fricke J, Kavanagh DG, Irvine DJ. In vitro and in vivo mRNA delivery using lipid-enveloped pH-responsive polymer nanoparticles. Molecular Pharmaceutics, 2011; 8(3): 774–787.

71. Kafshgari MH, Alnakhli M, Delalat B, Apostolou S, Harding FJ, Mäkilä E, Salonen JJ, Kuss BJ, Voelcker NH. Small interfering RNA delivery by polyethylenimine-functionalised porous silicon nanoparticles. Biomaterials Science, 2015; 3(12): 1555–1565.

72. Albanese A, Tang PS, Chan WC. The effect of nanoparticle size, shape, and surface chemistry on biological systems. Annual Review of Biomedical Engineering, 2012; 14(1): 1–6.

73. Golestanipour A, Nikkhah M, Aalami A, Hosseinkhani S. Gene delivery to tobacco root cells with single-walled carbon nanotubes and cell-penetrating fusogenic peptides. Molecular Biotechnology, 2018; 60(12): 863–878.

74. Gao C. Genome engineering for crop improvement and future agriculture. Cell, 2021; 184(6): 1621–1635.

75. Hamada H, Linghu Q, Nagira Y, Miki R, Taoka N, Imai R. An in planta biolistic method for stable wheat transformation. Scientific Reports, 2017; 7(1): 1–8.

76. Imai R, Hamada H, Liu Y, Linghu Q, Kumagai Y, Nagira Y, Miki R, Taoka N. In planta particle bombardment (iPB): a new method for plant transformation and genome editing. Plant Biotechnology, 2020; 20–0206.

77. Cha C, Shin SR, Annabi N, Dokmeci MR, Khademhosseini A. Carbon-based nanomaterials: multifunctional materials for biomedical engineering. ACS Nano, 2013; 7(4): 2891–2897.

78. Majeed N, Panigrahi KC, Sukla LB, John R, Panigrahy M. Application of carbon nanomaterials in plant biotechnology. Materials Today: Proceedings, 2020; 30: 340–345.

79. Xu X, Ray R, Gu Y, Ploehn HJ, Gearheart L, Raker K, Scrivens WA. Electrophoretic analysis and purification of fluorescent single-walled carbon nanotube fragments. Journal of the American Chemical Society, 2004; 126(40): 12736–12737.

80. Misra SK, Srivastava I, Tripathi I, Daza E, Ostadhossein F, Pan D. Macromolecularly "caged" carbon nanoparticles for intracellular trafficking via switchable photoluminescence. Journal of the American Chemical Society, 2017; 139(5): 1746–1749.

81. Kim TH, Sirdaarta JP, Zhang Q, Eftekhari E, St John J, Kennedy D, Cock IE, Li Q. Selective toxicity of hydroxyl-rich carbon nanodots for cancer research. Nano Research, 2018; 11(4): 2204–2216.

82. Fernando KS, Sahu S, Liu Y, Lewis WK, Guliants EA, Jafariyan A, Wang P, Bunker CE, Sun YP. Carbon quantum dots and applications in photocatalytic energy conversion. ACS Applied Materials & Interfaces, 2015; 7(16): 8363–8376.

83. Gan Z, Wu X, Zhou G, Shen J, Chu PK. Is there real upconversion photoluminescence from graphene quantum dots? Advanced Optical Materials, 2013; 1(8): 554–558.

84. Baker SN, Baker GA. Luminescent carbon nanodots: emergent nanolights. Angewandte Chemie International Edition, 2010; 49(38): 6726–6744.

85. Schwartz SH, Hendrix B, Hoffer P, Sanders RA, Zheng W. Carbon dots for efficient small interfering RNA delivery and gene silencing in plants. Plant Physiology, 2020; 184(2): 647–657.

86. Eatemadi A, Daraee H, Karimkhanloo H, Kouhi M, Zarghami N, Akbarzadeh A, Abasi M, Hanifehpour Y, Joo SW. Carbon nanotubes: properties, synthesis, purification, and medical applications. Nanoscale Research Letters, 2014; 9(1): 1–3.

87. He H, Pham-Huy LA, Dramou P, Xiao D, Zuo P, Pham-Huy C. Carbon nanotubes: applications in pharmacy and medicine. BioMed Research International, 2013. doi.org/10.1155/2013/578290.

88. Beg S, Rizwan M, Sheikh AM, Hasnain MS, Anwer K, Kohli K. Advancement in carbon nanotubes: basics, biomedical applications and toxicity. Journal of Pharmacy and Pharmacology, 2011; 63(2): 141–163.

89. Schnorr JM, Swager TM. Emerging applications of carbon nanotubes. Chemistry of Materials, 2011; 23(3): 646–657.

90. Kang SJ, Kocabas C, Kim HS, Cao Q, Meitl MA, Khang DY, Rogers JA. Printed multilayer superstructures of aligned single-walled carbon nanotubes for electronic applications. Nano Letters, 2007; 7(11): 3343–3348.

91. Fu YQ, Li LH, Wang PW, Qu J, Fu YP, Wang H, Sun JR, Lü CL. Delivering DNA into plant cell by gene carriers of ZnS nanoparticles. Chemical Research in Chinese Universities, 2012; 28(4): 672–676.

92. Zheng M, Jagota A, Semke ED, Diner BA, McLean RS, Lustig SR, Richardson RE, Tassi NG. DNA-assisted dispersion and separation of carbon nanotubes. Nature Materials, 2003; 2(5): 338–342.

93. Vardharajula S, Ali SZ, Tiwari PM, Eroğlu E, Vig K, Dennis VA, Singh SR. Functionalized carbon nanotubes: biomedical applications. International Journal of Nanomedicine, 2012; 7: 5361.

94. Lombardo D, Kiselev MA, Caccamo MT. Smart nanoparticles for drug delivery application: development of versatile nanocarrier platforms in biotechnology and nanomedicine. Journal of Nanomaterials, 2019. doi.org/10.1155/2019/3702518.

95. Rastogi A, Tripathi DK, Yadav S, Chauhan DK, Živčák M, Ghorbanpour M, El-Sheery NI, Brestic M. Application of silicon nanoparticles in agriculture. 3 Biotech, 2019; 9(3): 1–1.

96. Tang F, Li L, Chen D. Mesoporous silica nanoparticles: synthesis, biocompatibility and drug delivery. Advanced Materials, 2012; 24(12): 1504–1534.

97. Zhou Y, Quan G, Wu Q, Zhang X, Niu B, Wu B, Huang Y, Pan X, Wu C. Mesoporous silica nanoparticles for drug and gene delivery. Actapharmaceutica Sinica B, 2018; 8(2): 165–177.

98. Kumar H, Venkatesh N, Bhowmik H, Kuila A. Metallic nanoparticle: a review. Biomedical Journal of Scientific & Technical Research, 2018; 4(2): 3765–3775.

99. McNamara K, Tofail SA. Nanoparticles in biomedical applications. Advances in Physics: X, 2017; 2(1): 54–88.

100. Sharma A, Goyal AK, Rath G. Recent advances in metal nanoparticles in cancer therapy. Journal of Drug Targeting, 2018; 26(8): 617–632.

101. Brown KR, Walter DG, Natan MJ. Seeding of colloidal Au nanoparticle solutions. Improved control of particle size and shape. Chemistry of Materials, 2000; 12(2): 306–313.

102. Hussain JI, Kumar S, Hashmi AA, Khan Z. Silver nanoparticles: preparation, characterization, and kinetics. Advanced Materials Letters, 2011; 2(3): 188–194.

103. Han G, Ghosh P, Rotello VM. Functionalized gold nanoparticles for drug delivery. Nanomedicine, 2007; 2(1): 113–123.
104. Nishiyama N. Nanomedicine: nanocarriers shape up for long life. Nature Nanotechnology, 2007; 2(4): 203.
105. Majidi S, ZeinaliSehrig F, Samiei M, Milani M, Abbasi E, Dadashzadeh K, Akbarzadeh A. Magnetic nanoparticles: applications in gene delivery and gene therapy. Artificial Cells, Nanomedicine, and Biotechnology, 2016; 44(4): 1186–1193.
106. Kudr J, Haddad Y, Richtera L, Heger Z, Cernak M, Adam V, Zitka O. Magnetic nanoparticles: from design and synthesis to real world applications. Nanomaterials, 2017; 7(9): 243.
107. Hao Y, Yang X, Shi Y, Song S, Xing J, Marowitch J, Chen J, Chen J. Magnetic gold nanoparticles as a vehicle for fluorescein isothiocyanate and DNA delivery into plant cells. Botany, 2013; 91(7): 457–466.
108. Berry CC. Progress in functionalization of magnetic nanoparticles for applications in biomedicine. Journal of Physics D: Applied Physics, 2009; 42(22): 224003.
109. McBain SC, Yiu HH, Dobson J. Magnetic nanoparticles for gene and drug delivery. International Journal of Nanomedicine, 2008; 3(2): 169.
110. Choy JH, Kwak SY, Jeong YJ, Park JS. Inorganic layered double hydroxides as nonviral vectors. Angewandte Chemie International Edition, 2000; 39(22): 4041–4045.
111. Li P, Duan X, Kuang Y, Li Y, Zhang G, Liu W, Sun X. Tuning electronic structure of NiFe layered double hydroxides with vanadium doping toward high efficient electrocatalytic water oxidation. Advanced Energy Materials, 2018; 8(15): 1703341.
112. Fan G, Li F, Evans DG, Duan X. Catalytic applications of layered double hydroxides: recent advances and perspectives. Chemical Society Reviews, 2014; 43(20): 7040–7066.
113. Li X, Du D, Zhang Y, Xing W, Xue Q, Yan Z. Layered double hydroxides toward high-performance supercapacitors. Journal of Materials Chemistry A, 2017; 5(30): 15460–15485.
114. Sajid M, Basheer C. Layered double hydroxides: emerging sorbent materials for analytical extractions. TrAC Trends in Analytical Chemistry, 2016; 75: 174–182.
115. Nummelin S, Kommeri J, Kostiainen MA, Linko V. Evolution of structural DNA nanotechnology. Advanced Materials, 2018; 30(24): 1703721.
116. Wang Y, Benson E, Fördős F, Lolaico M, Baars I, Fang T, Teixeira AI, Högberg B. DNA origami penetration in cell spheroid tissue models is enhanced by wireframe design. Advanced Materials, 2021; 33(29): 2008457.
117. Zhang H, Zhang H, Demirer GS, Gonzalez-Grandio E, Fan C, Landry MP. Engineering DNA nanostructures for siRNA delivery in plants. Nature Protocols, 2020; 15(9): 3064–3087.
118. Wang F, Liu B, Ip AC, Liu J. Orthogonal adsorption onto nano-graphene oxide using different intermolecular forces for multiplexed delivery. Advanced Materials, 2013; 25(30): 4087–4092.
119. Caboche M. Liposome-mediated transfer of nucleic acids in plant protoplasts. Physiologia Plantarum, 1990; 79(1): 173–176.
120. Lurquin PF, Rollo F. Liposome-mediated delivery of nucleic acids into plant protoplasts. In Methods in Enzymology. Vol. 221. Elsevier, Amsterdam, 1993; 409–415.
121. Ahokas H. Transfection by DNA-associated liposomes evidenced at pea pollination. Hereditas, 1987; 106(1): 129–138.
122. Tuncel D, Demir HV. Conjugated polymer nanoparticles. Nanoscale, 2010; 2(4): 484–494.

123. Divya K, Jisha MS. Chitosan nanoparticles preparation and applications. Environmental Chemistry Letters, 2018; 16(1): 101–112.
124. Abdel-Aziz HM, Hasaneen MN, Omer AM. Nano chitosan-NPK fertilizer enhances the growth and productivity of wheat plants grown in sandy soil. Spanish Journal of Agricultural Research, 2016; 14(1): e0902–e0902.
125. Zhao X, Yu SB, Wu FL, Mao ZB, Yu CL. Transfection of primary chondrocytes using chitosan-pEGFP nanoparticles. Journal of Controlled Release, 2006; 112(2): 223–228.
126. Grenha A, Grainger CI, Dailey LA, Seijo B, Martin GP, Remuñán-López C, Forbes B. Chitosan nanoparticles are compatible with respiratory epithelial cells in vitro. European Journal of Pharmaceutical Sciences, 2007; 31(2): 73–84.
127. Guo L, Yan DD, Yang D, Li Y, Wang X, Zalewski O, Yan B, Lu W. Combinatorial photothermal and immuno cancer therapy using chitosan-coated hollow copper sulfide nanoparticles. ACS Nano, 2014; 8(6): 5670–5681.
128. Araújo BR, Romao LP, Doumer ME, Mangrich AS. Evaluation of the interactions between chitosan and humics in media for the controlled release of nitrogen fertilizer. Journal of Environmental Management, 2017; 190: 122–131.
129. Amar-Lewis E, Azagury A, Chintakunta R, Goldbart R, Traitel T, Prestwood J, Landesman-Milo D, Peer D, Kost J. Quaternized starch-based carrier for siRNA delivery: from cellular uptake to gene silencing. Journal of Controlled Release, 2014; 185: 109–120.
130. Hoyer JA, Neundorf I. Peptide vectors for the nonviral delivery of nucleic acids. Accounts of Chemical Research, 2012; 45(7): 1048–1056.
131. Fonseca SB, Pereira MP, Kelley SO. Recent advances in the use of cell-penetrating peptides for medical and biological applications. Advanced Drug Delivery Reviews, 2009; 61(11): 953–964.
132. Meng Z, Luan L, Kang Z, Feng S, Meng Q, Liu K. Histidine-enriched multifunctional peptide vectors with enhanced cellular uptake and endosomal escape for gene delivery. Journal of Materials Chemistry B, 2017; 5(1): 74–84.
133. Lakshmanan M, Kodama Y, Yoshizumi T, Sudesh K, Numata K. Rapid and efficient gene delivery into plant cells using designed peptide carriers. Bio-Macromolecules, 2013; 14(1): 10–16.
134. Su LX, Ma XL, Zhao KK, Shen CL, Lou Q, Yin DM, Shan CX. Carbon nanodots for enhancing the stress resistance of peanut plants. ACS Omega, 2018; 3(12): 17770–17777.
135. Szabó A, Perri C, Csató A, Giordano G, Vuono D, Nagy JB. Synthesis methods of carbon nanotubes and related materials. Materials, 2010; 3(5): 3092–3140.
136. Nandy B, Santosh M, Maiti PK. Interaction of nucleic acids with carbon nanotubes and dendrimers. Journal of Biosciences, 2012; 37(3): 457–474.
137. Chen Y, Liu H, Ye T, Kim J, Mao C. DNA-directed assembly of single-wall carbon nanotubes. Journal of the American Chemical Society, 2007; 129(28): 8696–8697.
138. Demirer GS, Zhang H, Goh NS, Pinals RL, Chang R, Landry MP. Carbon nanocarriers deliver siRNA to intact plant cells for efficient gene knockdown. Science Advances, 2020; 6(26): eaaz0495.
139. Wong MH, Misra RP, Giraldo JP, Kwak SY, Son Y, Landry MP, Swan JW, Blankschtein D, Strano MS. Lipid exchange envelope penetration (LEEP) of nanoparticles for plant engineering: a universal localization mechanism. Nano Letters, 2016; 16(2): 1161–1172.
140. Kesse S, Boakye-Yiadom KO, Ochete BO, Opoku-Damoah Y, Akhtar F, Filli MS, Asim Farooq M, Aquib M, MaviahMily BJ, Murtaza G, Wang B. Mesoporous silica

nanomaterials: versatile nanocarriers for cancer theranostics and drug and gene delivery. Pharmaceutics, 2019; 11(2): 77.

141. Slowing II, Trewyn BG, Lin VS. Mesoporous silica nanoparticles for intracellular delivery of membrane-impermeable proteins. Journal of the American Chemical Society, 2007; 129(28): 8845–8849.

142. Hussain HI, Yi Z, Rookes JE, Kong LX, Cahill DM. Mesoporous silica nanoparticles as a biomolecule delivery vehicle in plants. Journal of Nanoparticle Research, 2013; 15(6): 1–5.

143. Torney F, Trewyn BG, Lin VS, Wang K. Mesoporous silica nanoparticles deliver DNA and chemicals into plants. Nature Nanotechnology, 2007; 2(5): 295–300.

144. Martin-Ortigosa S, Valenstein JS, Lin VS, Trewyn BG, Wang K. Gold functionalized mesoporous silica nanoparticle mediated protein and DNA codelivery to plant cells via the biolistic method. Advanced Functional Materials, 2012; 22(17): 3576–3582.

145. Hajiahmadi Z, Shirzadian-Khorramabad R, Kazemzad M, Sohani MM. Enhancement of tomato resistance to *Tuta absoluta* using a new efficient mesoporous silica nanoparticle-mediated plant transient gene expression approach. Scientia Horticulturae, 2019; 243: 367–375.

146. Vijayakumar PS, Abhilash OU, Khan BM, Prasad BL. Nanogold-loaded sharp-edged carbon bullets as plant-gene carriers. Advanced Functional Materials, 2010; 20(15): 2416–2423.

147. Fu YQ, Li LH, Wang PW, Qu J, Fu YP, Wang H, Sun JR, Lü CL. Delivering DNA into plant cell by gene carriers of ZnS nanoparticles. Chemical Research in Chinese Universities, 2012; 28(4): 672–676.

148. Mitter N, Worrall EA, Robinson KE, Li P, Jain RG, Taochy C, Fletcher SJ, Carroll BJ, Lu GQ, Xu ZP. Clay nanosheets for topical delivery of RNAi for sustained protection against plant viruses. Nature Plants, 2017; 3: 16207. doi.org/10.1038/nplants.2016.207.

149. Karny A, Zinger A, Kajal A, Shainsky-Roitman J, Schroeder A. Therapeutic nanoparticles penetrate leaves and deliver nutrients to agricultural crops. Scientific Reports, 2018; 8: 7589. doi.org/10.1038/s41598-018-25197-y.

150. Nagata T, Okada K, Takebe I, Matsui C. Delivery of tobacco mosaic virus RNA into plant protoplasts mediated by reverse-phase evaporation vesicles (liposomes). Molecular and General Genetics MGG, 1981; 184(2): 161–165.

151. Rosenberg N, Dekel-Reichenbach M, Navot N, Gad AE, Altman A, Czosnek H. Liposome-mediated introduction of DNA into plant protoplasts and calli. International Symposium on In Vitro Culture and Horticultural Breeding, 1989; 30: 509–516.

152. Demirer GS, Landry MP. Delivering genes to plants. Chemical Engineering Progress, 2017; 113(4): 40–45.

153. Nandy D, Maity A, Mitra AK. Target-specific gene delivery in plant systems and their expression: insights into recent developments. Journal of Biosciences, 2020; 45(1): 1–2.

154. Cosgrove DJ. Growth of the plant cell wall. Nature Reviews Molecular Cell Biology, 2005; 6(11): 850–861.

155. Carpita N, Sabularse D, Montezinos D, Delmer DP. Determination of the pore size of cell walls of living plant cells. Science, 1979; 205(4411): 1144–1147.

156. Deng H, Huang W, Zhang Z. Nanotechnology based CRISPR/Cas9 system delivery for genome editing: progress and prospect. Nano Research, 2019; 12(10): 2437–2450.

157. Landry MP, Mitter N. How nanocarriers delivering cargos in plants can change the GMO landscape. Nature Nanotechnology, 2019; 14(6): 512–514.

158. Cox A, Venkatachalam P, Sahi S, Sharma N. Silver and titanium dioxide nanoparticle toxicity in plants: a review of current research. Plant Physiology and Biochemistry, 2016; 107: 147–163.
159. Du W, Tan W, Peralta-Videa JR, Gardea-Torresdey JL, Ji R, Yin Y, Guo H. Interaction of metal oxide nanoparticles with higher terrestrial plants: physiological and biochemical aspects. Plant Physiology and Biochemistry, 2017; 110: 210–225.
160. Rastogi A, Zivcak M, Sytar O, Kalaji HM, He X, Mbarki S, Brestic M. Impact of metal and metal oxide nanoparticles on plant: a critical review. Frontiers in Chemistry, 2017; 5: 78.
161. Dutta Gupta S, Saha N, Agarwal A, Venkatesh V. Silver nanoparticles (AgNPs) induced impairment of in vitro pollen performance of *Peltophorum pterocarpum* (DC.) K. Heyne. Ecotoxicology, 2020; 29(1): 75–85.
162. Hossain Z, Mustafa G, Komatsu S. Plant responses to nanoparticle stress. International Journal of Molecular Sciences, 2015; 16(11): 26644–26653.
163. Saptarshi SR, Duschl A, Lopata AL. Interaction of nanoparticles with proteins: relation to bio-reactivity of the nanoparticle. Journal of Nanobiotechnology, 2013; 11(1): 1–2.
164. Fleischer CC, Payne CK. Nanoparticle–cell interactions: molecular structure of the protein corona and cellular outcomes. Accounts of Chemical Research, 2014; 47(8): 2651–2659.
165. Ishida Y, Hiei Y, Komari T. *Agrobacterium*-mediated transformation of maize. Nature Protocols, 2007; 2(7): 1614–1621.
166. Mohammed S, AbdSamad A, Rahmat Z. *Agrobacterium*-mediated transformation of rice: constraints and possible solutions. Rice Science, 2019; 26(3): 133–146.
167. Zaidi SS, Mansoor S. Viral vectors for plant genome engineering. Frontiers in Plant Science, 2017; 8: 539.
168. Wang K, Frame B. Biolistic gun-mediated maize genetic transformation. *In*: Transgenic Maize. Humana Press, Totowa, NJ. 2009; 29–45.
169. Wright M, Dawson J, Dunder E, Suttie J, Reed J, Kramer C, Chang Y, Novitzky R, Wang H, Artim-Moore L. Efficient biolistic transformation of maize (*Zea mays* L.) and wheat (*Triticum aestivum* L.) using the phosphomannose isomerase gene, *pmi*, as the selectable marker. Plant Cell Reports, 2001; 20(5): 429–436.
170. Jones MG. Electroporation-mediated gene transfer to protoplasts and regeneration of transgenic plants. *In*: Gene Transfer to Plants. Springer, Berlin, 1995; 83–92.
171. Arencibia A, Molina P, Gutierrez C, Fuentes A, Greenidge V, Menendez E, De la Riva G, Selman-Housein G. Regeneration of transgenic sugarcane (*Saccharum officinarum* L.) plants from intact meristematic tissue transformed by electroporation. Biotecnologia Aplicada, 1992; 9(2): 156–165.

7 Nanodiagnostics
Tool for Diagnosis of Plant Pathogens

Gaurav Verma, Krishna Kant Mishra,
Baswaraj Raigond, Jeevan Bettanayaka,
Ashish Kumar Singh, Amit Umesh Paschapur,
and Lakshmi Kant

CONTENTS

7.1 INTRODUCTION

The ability to identify and diagnose a disease early is attributable to nanotechnology which has a considerable influence on disease diagnosis. Nanomaterials are widely used in the disciplines of organic, inorganic, and biological sciences, both structurally and functionally. These materials are exceptional and essential in many sectors of agriculture due to their special size-dependent characteristics. Nanotechnology is very useful for diagnosing pathogens, and different nanoparticles have shown promising results for identifying disease biomarkers.

While pathogen detection and management costs remain low, the global spread of plant diseases has grown [1]. Global losses were estimated to be 14% due to insect pests, 13% due to plant diseases, and 13% due to weeds. Pathogens limit plant development and productivity by imposing persistent stress, and accurate disease detection necessitates costly diagnostic tools [2]. Numerous initiatives have been made to produce crops more safely in various conditions using safeguards or best management techniques [2]. However, crop protection is essential for ensuring sustainable crop output, particularly in challenging environmental circumstances. Nanotechnology is already beginning to have an influence on the diagnosis, treatment, and prevention of disease, particularly by enabling early disease diagnosis and management as well as precise and effective intervention [3]. Nanomaterials are widely used in the disciplines of organic, inorganic, and biological materials, both structurally and functionally. These materials are outstanding and essential in many facets of human endeavor due to their special size-dependent characteristics. Numerous advancements have already begun, and in the next few years, the field of diagnostics will clearly see their effects [4].

Around the world, traditional molecular diagnostic techniques are frequently utilized to identify plant pathogenic organisms with a high degree of sensitivity and specificity. However, the majority of these methods cannot be used in the field (on-site detection) or in low-income developing nations. Furthermore, the use of conventional molecular techniques in developing countries is constrained by the high cost and short shelf life of some molecular biology reagents, such as enzymes and primers. Numerous issues in agriculture, such as the control of plant diseases, may actually have solutions thanks to nanotechnology. The effectiveness of fungicides and insecticides will be improved by nano-based materials, allowing for the use of much lower quantities. Additionally, nanodiagnostics and microfluidics provide innovative tools to enhance the sample preparation stage, which is still challenging to incorporate in a small platform. The target amplification strategies may need to cope with the signal amplification ones. The agricultural and food industries will also benefit from quick on-site diagnosis of plant infections using nano-based kits, nanosensors, nanobiosensors, nanobarcodes, and other portable diagnostic systems. The principles and current state of the application of nanotechnology in plant pathology are discussed in this review, along with nanodiagnosis methods using portable polymerase chain reaction (PCR) systems, nanopore sequencing tools, nanodiagnostic kits, gold nanoparticles, quantum dots (QDs), nanobarcodes, and nanosensors.

One of the most fascinating and dynamic fields of research, nanotechnology has a significant impact on a wide range of fields, including science, engineering, medicine, and agriculture. Nanomaterials, which typically have diameters between 1 and 100 nm, can offer increased surface-to-volume ratios as well as special chemical, optical, and electrical properties that cannot be found in their bulk equivalents, making them excellent candidates for the analysis of plant diseases. Due to their small sizes and quick diffusion rates, nanoscale materials can also interact with biomolecular targets more effectively.

Nanomaterials can be created in a wide variety of morphologies, such as spherical particles, cubes, rods, wires, plates, prisms, core shell structures, and more complex three-dimensional architectures. These morphologies can change shape or aggregate, and this can change a nanomaterial's chemical or physical properties in response to various environmental factors. In essence, this turns into one of the most typical nanomaterial sensing techniques. Developments in nanotechnology make it possible to create nanoparticles for a variety of biosensors or bioimaging applications, particularly those for monitoring patient care. The typical molecular detection toolkit now includes a wide variety of micro- or nanostructured platforms or sensors that offer exceptional sensitivity and spatiotemporal resolution. The domain of agricultural biosensors has also started to flourish in recent years. Nanosensors have already shown great promise for the reliable detection of pathogens that cause serious infections of a variety of commercially significant crops and plants due to their low cost and portability in the field.

7.2 TRADITIONAL DIAGNOSTIC SYSTEM FOR PLANT PATHOGENS

Monitoring plant health and implementing an effective integrated disease management strategy depend on the early detection of plant pathogens. Differentiating between causative species is crucial because numerous fungal infections alter plants in ways that are similar to one another during disease development. In susceptible crops, more obvious symptoms frequently manifest as morphological and color changes, particular necrotic patches, and even the loss of the plant's stem or leaves. To achieve completely informed management, understanding latent infection with no obvious signs is equally essential [5].

Visual crop inspection, which requires a skilled grower or pathologist, is the oldest traditional method that is still widely employed for disease and potentially pathogen diagnosis. When a visual diagnosis is achieved, the pathogen will probably have already been established in host populations. Therefore, much focus has been dedicated to the development of early pathogen detection methods with improved sensitivity, accuracy, and identification efficiency. Enzyme-linked immunosorbent assay (ELISA), PCR, and loop-mediated isothermal amplification (LAMP) tests have been the three main types of serological and molecular assays used up to this point, all of which are protein- or nucleic acid-based technologies.

Antibodies designed to recognize target-specific antigens can be used to identify and, to a lesser extent, quantify plant pathogens. ELISA is by far the most advanced serology-based diagnostics approach for viruses and fungi. This makes it possible to detect pathogens using a colorimetric reaction that may be seen with the unaided eye or measured with an optical reader [6]. *Botrytis* fungus species that causes *Botrytis* gray mold have been identified using colorimetric assays with monoclonal antibodies [7].

Traditional ELISA still has very low sensitivity and accuracy, even though it has almost universally been adopted as the gold standard for the identification of infections in environmental, chemical, biotechnological, health, and agricultural analyses. The major ELISA component, enzyme, which serves as a signal label and catalyst, could be replaced with nanoparticles. The changes significantly increase the sensitivity. Although traditional ELISA has improved significantly as a result of the incorporation of nanomaterials, there are still several aspects of this technology that could be improved.

For greater specificity and sensitivity of detection, other techniques that focus on the nucleic acid sequences of target pathogens have been developed. These frequently involve amplicon detection and visualization after amplicon detection, which amplifies the target sequence using PCR. Applications for pathogen diagnostics have adapted a range of PCR amplification assay techniques to improve the sensitivity, signal detection, and multiplexing capabilities [8]. Pathogen diagnostic applications have adopted a variety of PCR amplification assay types to enhance the assay's sensitivity (nested PCR), signal detection (magnetic capture hybridization and ELISA-PCR), and multiplexing capability (multiplex PCR) [8]. The PCR-based probes frequently target the ribosomal intergenic spacer [9], internal transcribed spacer regions, or house-keeping genes like β-tubulin [10]. Simple, dependable, scalable PCR techniques may identify picogram levels of pure fungus DNA [11]. To increase the usability of molecular diagnostic tests, LAMP, a DNA amplification technique, was developed. LAMP quickly amplifies nucleic acids with excellent specificity and sensitivity under isothermal conditions [12]. LAMP-diagnostic assays have recently been utilized to identify plant fungal pathogens in addition to bacterial, viral, and viroid pathogens [13–16].

Despite gains in sensitivity and specificity for a given target pathogen, these widely used approaches still have certain drawbacks, including longer diagnosis times, difficult sample preparation processes, transferring samples from the field to sophisticated labs, and the requirement for trained personnel. It is therefore necessary to develop an in-field diagnostic procedure that is quicker, more dependable, more sensitive, and accurate. Based on the primary properties of optics or electrochemistry, a "point-of-care technology" might be created. With its quick reaction time, cheap cost on-site trials, and lack of user experience in data interpretation, this method improves several bioassays [17].

7.3   BIOSENSORS IN PLANT DISEASE DIAGNOSIS

The diagnostic serves as the foundation for managing plant diseases and predicting crop loss due to plant pathogen infection. Early detection of infectious agent is the key factor affecting a crop's productivity. Most of the time, pesticides are used as a preventive measure, which increases the crop's residual toxicity as well as the risk to the environment. On the other hand, applying pesticides after the detection of the disease has manifested results in an additional financial expenditure in crop management. Due to the focus on preventing their vectors and, by extension, their spread, viral infections are the most challenging to manage. Therefore, identifying the disease during early detection, such as the replication of viral DNA or the beginning creation of its viral protein, is the key to success in the control based on a crop management system, particularly those of viral origin [18, 19]. Biosensors enable quick, efficient, and cost-effective detection of a specified analyte. This method, which can serve as an alternative to current detection systems, uses antibodies or antigens connected to a transducer that produces an analytical response. Due to its quick response time, user-friendly application, affordable production, and potential for system miniaturization, electrochemical transduction is one of the most common modalities used today. An electrochemical biosensor's fundamental function is to track changes that occur close to the electrode surface. The electrochemical immunosensors among them exhibit greater potential due to their increased sensitivity, increased speed, and constant control [20]. A microfluidic electrochemical immunosensor was developed for early detection of *Xanthomonas arboricola* in walnut plant samples [21]. This *in-situ* diagnosis offered much greater specificity and sensitivity than ELISA. A portable electromechanical immunosensor device based on immobilized cucumber mosaic virus

(CMV)-specific antibodies coated with gold nanoparticles (AuNPs) was performed for the detection of CMV. With the use of immunosensors, real-time and accurate CMV detection in nanophytopathology is now possible [22]. These biosensors are inexpensive, easy to use, and quickly identify target pathogens in the field for the detection of plant disease. The biosensors are also very sensitive and precise. For instance, a nanoparticle electrochemical biosensor was more sensitive than traditional PCR in detecting *Pseudomonas syringae*, and it was able to identify infected plants before any disease signs manifested [23]. To identify three fungal plant diseases concurrently (*Botrytis cinerea*, *Didymella bryoniae*, and *Botrytis squamosa*), Wang and Li devised the microfluidic microarray construction technique [24]. The advantage of the method includes quicker DNA hybridization and DNA probe creation with minimal sample requirement [24].

7.4 DETECTION OF FUNGAL TOXINS

Mycotoxins are secondary metabolites produced by fungi that grow on plant foods or animal fodder as parasites or saprophytes. They mostly consist of *Aspergillus*, *Penicillium*, and *Fusarium* individuals. Worldwide, people rely heavily on grains, which are frequently contaminated with fungus that creates mycotoxins. Mycotoxin contamination of food crops and goods can have severe financial and societal repercussions. The examination of mycotoxin uses a variety of approaches. These primarily consist of immunoaffinity columns, screening cards, lateral flow tests, and ELISA tests. Recently, the quick detection of mycotoxins has been made possible by the use of nanotechnology in biosensors. Glassy carbon electrodes with carbon dots were used to measure how the voltammetric behavior of the curcumin-containing metanil yellow changed [25]. Electrochemical biosensors that are dependable, sophisticated, sensitive, and reasonably priced are essential for detecting food toxins [26]. Mycotoxins were detected using electrochemical biosensors based on nanomaterials. Recent tests on glassy carbon electrodes covered with calixarene and AuNPs revealed the presence of metanil yellow and fast green as well. Utilizing electrochemical impedance spectroscopy, cyclic voltammetry, and differential pulse voltammetry, the sensitivity was determined [27].

7.5 QUANTUM DOTS

QDs are inorganic fluorophores with significant benefits over standard organic fluorophores employed as visual detection markers on nucleic acids or proteins, and these nanomaterials have been most commonly used for disease diagnostics [28–30]. Most frequently, viral nucleic acids have been utilized in hybridization assays and immunoassays as a technique to allow Förster resonance energy transfer (FRET) after their detection, as well as antigen–antibody affinity. It is a non-radiative energy transfer between donor and acceptor fluorophores [31]. *Citrus tristeza* virus (CTV) nanodiagnostic was reported in which AuNPs were coated with cadmium telluride (CdTe) QDs conjugated with a specific antibody against the CTV coat protein to create a focused and sensitive FRET-based nanobiosensor [32]. Various bacteria have also been employed to biosynthesize cadmium sulfide; however, few investigations have concentrated on its luminous characteristics. When treated with a combination of $CdCl_2$ and $TeCl_2$ at ambient conditions, the *Fusarium oxysporum* isolates successfully produced highly fluorescent CdTe QDs by myco-mediated synthesis [33]. Selected-area electron diffraction and transmission electron microscopy were used to describe these biosynthesized CdTe nanoparticles [34].

7.6 GOLD NANOPARTICLES

Because of their optical and electrical properties, compatibility, and ease of modification, AuNPs are employed in biosensors. The unique colorimetric or optical detecting capabilities offered by AuNPs make their use in biosensors exceptional. One such area where amazing advancements have been made is the biosensing of disease-causing pathogens in plants utilizing nanomaterials [35]. In order to identify the presence of the disease rapidly and on-site, single-stranded DNA (ssDNA) probes specific to *Acidovorax avenae* subspecies *citrulli*, the bacterial pathogen that causes fruit blotch, were labelled with colloidal AuNPs [36]. Another example of the detection of *Pseudomonas syringae* pathovars, which cause prevalent bacterial infections in agricultural plants, used AuNP-tagged DNA probes in a colorimetric detection of pathogen DNA molecules. The *hrcV* gene's conserved N-terminal region served as the basis for the creation of specific primers, and the 5' or 3' ends of the probes were built with thiol capping. A color change from red (non-hybridized) to purple (probe-hybridized) indicated the presence of pathogen DNA in the sample via colorimetric detection [37]. Transmission electron microscopy has traditionally been the major method for identifying biospecific interactions with colloidal gold particles [38]. In addition to the typical colloidal gold made up of quasi-spherical particles (nanospheres), non-spherical particles can be employed, including nanorods, nanoshells, nanocages, and nanostars [39].

7.7 NANOPORE SEQUENCING

For both clinical and agricultural investigations, the detection accuracy and sensitivity of probing DNA structural variants have seen significant improvements because of the advent of single-molecule sequencing technology [40]. The nanopore sequencing platform benefits from real-time and lengthy reads of sequence data, quick running times, scalable genome mapping, and modest sample loadings as compared to preceding generations of sequencing techniques [41]. The diagnosis of plant pathogens has been a major focus of several recent investigations employing nanopore sequencing technologies. For instance, an effective standard protocol for the diagnosis of various plant bacteria, viruses, fungi, and phytoplasma, such as *Penicillium digitatum* in lemon and *Solanum lycopersicum* in tomato, using a handheld sequencing system, was created by Oxford Nanopore Technologies (MinION) [42]. The entire assay takes less than 2 hours, and the outcomes are equivalent to those of traditional diagnostic techniques (e.g., PCR and ELISA). The nanopore technology might enable diagnostic devices now in use to analyze the complete genome in a matter of minutes as opposed to hours.

7.8 NANOCHIP IN PLANT DISEASE DIAGNOSIS

In most cases, a NanoChip assay is built on the hybridization of complementary DNA strands using oligo capture probes, which also makes it possible for multiplex assays to detect the nucleotide changes of the intended organism. This technique has the unique ability to use amplified PCR samples or oligo capture probes that have been biotinylated and immobilized [43, 44]. NanoChips have demonstrated great specificity and accuracy to diagnose potato virus Y (PVY), potato virus X (PVX), and potato leafroll virus (PLRV) in potato due to their capacity to detect single nucleotide alterations. The Nanogen system also makes it possible to distinguish between the PVY, PVY⁰, and PVYN/NTN races [45]. A timely, precise, and effective diagnosis is crucial in the management of viral infections since preventive control only works when the crop is virus-free. In this sense, making a diagnosis solely based on symptoms is ineffective because they can also be caused by nutritional deficiencies, host–virus interactions, and co-infections. The various nanodiagnostic technologies can therefore be used depending on the goal of the research. According to this concept, research must be done in order to improve these tests in order to get accurate findings when the pathogen is present at low titers in its host.

7.9 NANOPROBE-BASED LOOP-MEDIATED ISOTHERMAL AMPLIFICATION

A set of six oligonucleotide primers with eight binding sites that specifically hybridize to various regions of a target gene and a thermophilic DNA polymerase from *Geobacillus stearothermophilus* are used in the LAMP method to detect genomic DNA specifically and quickly. By integrating various-sized QDs into polymeric microbeads, it has been possible to create AuNPs that are marked with DNA for use in biological assays [33]. Positive responses should primarily flash bright green, but negative reactions should continue to be pale orange. When SYBR Green I dye is applied, this would be the anticipated outcome, indicating that the LAMP amplicons would be immediately visible and observed with the naked eye or under ultraviolet transillumination.

7.10 BIO-BARCODE ASSAY (BIO-BARCODED DNA)

An incredibly sensitive technique for amplifying and finding proteins or nucleic acids is the bio-barcode test. DNA bio-barcoded assays use magnetic AuNPs customized with oligonucleotides for signal amplification and for the quick separation of a target protein from the sample. It is especially promising under ideal conditions; it enables the rapid detection of a wide range of protein targets at low-attomolar concentrations [46] and nucleic acids at high-zeptomolar levels [47]. The bio-barcode test is a novel idea that offers a potential substitution for the PCR method.

7.11 CONCLUSION

The portable diagnostic equipment, QDs, and bio-barcoded DNA sensor all have potential applications in the detection of plant diseases and mycotoxins of fungi.

Depending on the goal of the research, these nanodiagnostic technologies can be used for the proper management of the disease. The development of biochip technology will continue to improve future possibilities in plant disease diagnostics. Nanophytopathology can be used to understand better plant–pathogen interactions, perhaps leading to novel crop protection techniques.

REFERENCES

1. Agrios GN. Plant Pathology. 5th ed. San Diego, CA: Elsevier-Academic Press; 2005; p. 922.
2. Leake JR, Donnelly DP, Boddy L. Interactions between ecto-mycorrhizal fungi and saprotrophic fungi. In: Van der Heijden MGA, Sanders IR, editors. Mycorrhizal Ecology. Heidelberg: Ecological Studies Springer Verlag; 2002; p. 157.
3. Buxton DB, Lee SC, Wickline SA. Recommendations of the National Heart, Lung, and Blood Institute Nanotechnology Working Group. Circulation, 2003; 108(22): 2737–2742.
4. Yezhelyev MV, Gao X, Xing Y. Emerging use of nanoparticles in diagnosis and treatment of breast cancer. Lancet, 2006; 7(8): 657–667.
5. Oerke E-C. Remote sensing of diseases. Annual Review of Phytopathology, 2020; 58: 225–252.
6. Fang Y, Ramasamy R. Current and prospective methods for plant disease detection. Biosensors, 2015; 5: 537–561.
7. Dewey FM, Yohalem D. Detection, quantification and immunolocalisation of *Botrytis* species. In: Botrytis: Biology, Pathology and Control. Berlin, Germany: Springer; 2007; pp. 17–34.
8. Capote N, Pastrana AM, Aguado A, Sánchez-Torres P. Molecular tools for detection of plant pathogenic fungi and fungicide resistance. Plant Pathology, 2012; 59: 151–202.
9. Suarez MB, Walsh K, Boonham N, O'Neill T, Pearson S, Barker I. Development of real-time PCR (TaqMan) assays for the detection and quantification of *Botrytis cinerea* in planta. Plant Physiology and Biochemistry, 2005; 43: 890–899.
10. Brouwer M, Lievens B, Hemelrijck W, Ackerveken G, Cammue Thomma BPA, Thomma BPHJ. Quantification of disease progression of several microbial pathogens on *Arabidopsis thaliana* using real-time fluorescence PCR. FEMS Microbiology Letters, 2003; 228: 241–248.
11. Nielsen K, Yohalem DS, Jensen DF. PCR detection and RFLP differentiation of *Botrytis* species associated with neck rot of onion. Plant Disease, 2002; 86: 682–686.
12. Notomi T, Okayama H, Masubuchi H, Yonekawa T, Watanabe K, Amino N et al. Loop-mediated isothermal amplification of DNA. Nucleic Acids Research, 2000; 28: e63. doi:10.1093/nar/28.12.e63
13. Obura E, Masiga D, Wachira F, Gurja B, Khan ZR. Detection of phytoplasma by loop-mediated isothermal amplification of DNA (LAMP). Journal of Microbiological Methods, 2011; 84: 312–316. doi:10.1016/j.mimet.2010.12.011
14. Bühlmann A, Pothier JF, Rezzonico F, Smits THM, Andreou M, Boonham N et al. *Erwinia amylovora* loop-mediated isothermal amplification (LAMP) assay for rapid pathogen detection and on-site diagnosis of fire blight. Journal of Microbiological Methods, 2013; 92: 332–339. doi:10.1016/j.mimet.2012.12.017
15. Verma G, Raigond B, Pathania S et al. Development and comparison of reverse transcription-loop-mediated isothermal amplification assay (RT-LAMP), RT-PCR and

real time PCR for detection of potato spindle tuber viroid in potato. European Journal of Plant Pathology, 2020; 158: 951–964. https://doi.org/10.1007/s10658-020-02129-z

16. Manjunatha C, Sharma S, Kulshreshtha D, Gupta S, Singh K, Bhardwaj SC et al. Rapid detection of *Puccinia triticina* causing leaf rust of wheat by PCR and loop mediated isothermal amplification. PloS One, 2018; 13, e0196409. doi:10.1371/journal.pone.0196409

17. Cassedy A, Mullins E, O'Kennedy R. Sowing seeds for the future: the need for on-site plant diagnostics. Biotechnology Advances, 2020; 39: 107358. doi:10.1016/j.biotechadv.2019.02.014

18. Pearson MN, Clover GRG, Guy PL, Fletcher JD, Beever RE. A review of the plant virus, viroid and mollicute records for New Zealand. Australasian Plant Pathology, 2006; 35: 217–252.

19. Aboul-Ata AE, Mazyad H, El-Attar AK, Soliman AM, Anfoka G, Zeidaen M, Gorovits R, Sobol I, Czosnek H. Diagnosis and control of cereal viruses in the Middle East. Advances in Virus Research, 2011; 81: 33–61.

20. Ricci F, Volpe G, Micheli L, Palleschi G. A review on novel developments and applications of immunosensors in food analysis. Analytica Chimica Acta, 2007; 605: 111–127.

21. Regiart M, Fernández-Baldo MA, Villarroel-Rocha J, Messina GA, Bertolino FA, Sapag K et al. Microfluidic immunosensor based on mesoporous silica platform and CMK-3/poly-acrylamide-co-methacrylate of dihydrolipoic acid modified gold electrode for cancer biomarker detection. Analytica Chimica Acta, 2017; 963: 83–92. 10.1016/j.aca.2017.01.029

22. Rafidah AR, Faridah S, Shahrul AA, Mazidah M, Zamri I. Chronoamperometry measurement for rapid cucumber mosaic virus detection in plants. Proceedings of the Chemical Society, 2016; 20: 25–28.

23. Lau HY, Wu H, Wee EJH, Trau M, Wang Y, Botella JR. Specific and sensitive isothermal electrochemical biosensor for plant pathogen DNA detection with colloidal gold nanoparticles as probes. Scientific Reports, 2017; 7: 38896.

24. Wang L, Li PCH. Flexible microarray construction and fast DNA hybridization conducted on a microfluidic chip for greenhouse plant fungal pathogen detection. Journal of Agricultural and Food Chemistry, 2007; 55: 10509–10516.

25. Shereema RM, Rao TP, Sameer Kumar VB, Sruthi TV, Vishnu R, Prabhu GRD, Sharath Shankar S. Individual and simultaneous electrochemical determination of metanil yellow and curcumin on carbon quantum dots based glassy carbon electrode. Materials Science & Engineering C: Materials for Biological Applications, 2018; 93: 21–27. https://doi.org/10.1016/j.msec.2018.07.055.

26. Gupta R, Raza N, Bhardwaj SK, Vikrant K, Kim KH, Bhardwaj N. Advances in nanomaterial-based electrochemical biosensors for the detection of microbial toxins, pathogenic bacteria in food matrices. Journal of Hazardous Materials, 2021; 401. https://doi.org/10.1016/j.jhazmat.2020.123379.

27. Shah A. A novel electrochemical nanosensor for the simultaneous sensing of two toxic food dyes. ACS Omega, 2020; 5: 6187–6193. https://doi.org/10.1021/acsomega.0c00354.

28. Wang L, O'Donoghue MM, Tan W. Nanoparticles for multiplex diagnostics and imaging. Nanomedicine, 2006; 1(4): 413–426.

29. Adams FC, Barbante C. Nanoscience, nanotechnology and spectrometry. Spectrochimica Acta Part B, 2013; 86: 3–13.

30. Holzinger M, Le Goff A, Cosnier S. Nanomaterials for biosensing applications: a review. Frontiers in Chemistry, 2014; 2: 63.

31. Uhl J, Tang Y, Cockerill ER. Fluorescence resonance energy transfer. *In*: Persing D, Tenover F, Tang Y, Nolte F, Hayden R, van Belkum A (eds.) Molecular Microbiology. Washington, DC: ASM Press, 2011; pp. 231–244.

32. Shojaei TR, Salleh MAM, Sijam K, Rahim RA, Mohsenifar A, Safarnejad R, Tabatabaei M. Detection of *Citrus tristeza* virus by using fluorescence resonance energy transfer based biosensor. Spectrochimica Acta Part A: Molecular and Biomolecular Spectroscopy, 2016; 169: 216–222.

33. Jain K. Nanodiagnostics: application of nanotechnology (NT) in molecular diagnostics. Expert Review of Molecular Diagnostics, 2003; 2: 153–161.

34. Syed A, Ahmad A. Extracellular biosynthesis of CdTe quantum dots by the fungus *Fusarium oxysporum* and their anti-bacterial activity. Spectrochimica Acta Part A: Molecular and Biomolecular Spectroscopy, 2013; 106: 41–47.

35. Elmer W, White JC. The future of nanotechnology in plant pathology. Annual Review of Phytopathology, 2018; 56: 111–133.

36. Zhao W, Lu J, Ma W, Xu C, Kuang H, Zhu S. Rapid on-site detection of *Acidovorax avenae* subsp. *citrulli* by gold-labeled DNA strip sensor. Biosensors and Bioelectronics, 2011; 26: 4241–4244.

37. Vaseghi A, Safaie N, Bakhshinejad B, Mohsenifar A, Sadeghizadeh M. Detection of *Pseudomonas syringae* pathovars by thiol-linked DNA–gold nanoparticle probes. Sensors and Actuators B: Chemical, 2013; 181: 644–651. doi: 10.1016/j.snb.2013.02.018

38. Hayat MA. Colloidal gold: principles, methods, and applications, Vol. 1. San Diego, CA: Academic Press, 1989; p. 538.

39. Khlebtsov NG, Dykman LA. Optical properties and biomedical applications of plasmonic nanoparticles. Journal of Quantitative Spectroscopy and Radiative Transfer, 2010; 111: 1–35.

40. Bickhart DM, Rosen BD, Koren S, Sayre BL, Hastie AR, Chan S, Lee J, Lam ET, Liachko I, Sullivan ST, Burton JN, Huson HJ, Nystrom JC, Kelley M, Hutchison JL, Zhou Y, Sun J, Crisà A, Ponce de León FA, Schwartz JC, Hammond JA, Waldbieser GC, Schroeder SG, Liu GE, Dunham MJ, Shendure J, Sonstegard TS, Phillippy AM, Van Tassell CP, Smith TPL. Single-molecule sequencing and chromatin conformation capture enable de novo reference assembly of the domestic goat genome. Nature Genetics, 2017; 49: 643–650. https://doi.org/10.1038/ng.3802.

41. Watson CM, Crinnion LA, Hewitt S, Bates J, Robinson R, Carr IM, Sheridan E, Adlard J, Bonthron DT. Cas9-based enrichment and single-molecule sequencing for precise characterization of genomic duplications. Laboratory Investigation, 2020; 100(1): 135–146.

42. Chalupowicz L, Dombrovsky A, Gaba V, Luria N, Reuven M, Beerman A, Lachman O, Dror O, Nissan G, Manulis-Sasson S. Diagnosis of plant diseases using the nanopore sequencing platform. Plant Pathology, 2019; 68: 229–238.

43. Sosnowski RG, Tu E, Butler WF, O'Connell JP, Heller MJ. Rapid determination of single base mismatch mutations in DNA hybrids by direct electric field control. Proceedings of the National Academy of Science of the USA, 1997; 94: 1119–1123.

44. López MM, Llop P, Olmos A, Marco-Noales E, Cambra M, Bertolini E. Are molecular tools solving the challenges posed by detection of plant pathogenic bacteria and viruses? Current Issues in Molecular Biology, 2009; 11: 13–46.

45. Ruiz-García AB, Olmos A, Arahal DR, Antúnez O, Llop P, Pérez-Ortín JE, López MM, Cambra M. Biochip electrónico para la detección y caracterización simultánea de los principales virus y bacterias patógenos de la patata. XII Congreso de la Sociedad Española de Fitopatología, 2004; Lloret de Mar. 12 p.

46. Goluch ED, Nam JM, Georganopoulou DG, Chiesl TN, Shaikh KA, Ryu KS, Barron AE, Mirkin CA, Liu C. A bio-barcode assay for on-chip attomolar-sensitivity protein detection. Lab Chip, 2006; 6(10): 1293–1299.
47. Nam JM, Stoeva SI, Mirkin CA. Bio-bar-code-based DNA detection with PCR-like sensitivity. Journal of the American Chemical Society, 2004; 126(19): 5932–5933.

8 Myconanoparticles
Synthesis and Probable Role in Plant Pathogen Management

Mohadeseh Hassanisaadi, Mehrdad Chaichi,
Soheila Mirzaei, and Moslem Heydari

CONTENTS

DOI: 10.1201/9781003322122-8

8.1   INTRODUCTION

The technology under discussion about deals with materials production in a different knowledge group due to their very small size: "nanotechnology." The word "nano" comes from a unit of measurement meaning one billionth. In other words, in the metric system, it is 10^{-9} meters. Nanomaterials have a very small structure in the range of 1–100 nm, and this small size has differentiated their properties from bulk materials. These unique properties are related not only to their nanoscale size but also to their high surface area [1]. Their chemically reactive, optically active, and mechanically strong properties make them suitable for various applications [2]. Due to the extensive properties of nanomaterial, as mentioned earlier, these compounds are also widely used in different industries and have created profound and growing changes in various fields of science.

Nanoparticles (NPs) can be made by breaking down the desired bulk materials or they can be synthesized by various chemical or biological methods [3]. The top-to-bottom and bottom-to-up approaches refer to the above methods, respectively, which is quite a conceptual naming. However, the word "synthesis" is much more appropriate for the second method. The method of making NPs by breaking down bulk materials, in addition to the need for large equipment, requires a significant amount of energy [4]; furthermore, the surface structure of NPs is not sufficiently perfect in this method [5]. Since we know that many of the specific physical properties of NPs are due to their surface structure, their synthesis method is one of the building blocks of their properties. Unlike the previous method, the synthesis of nanomaterials through chemical systems is a fast and usable method for producing a wide range of nanomaterials. However, this method is not environmentally friendly and is associated with generating toxic and hazardous substances for the environment [6]. The last but newly developed method is the use of biological agents, which has attracted a great deal of attraction despite being at the beginning of its path. This is clearly due to its extremely ecologically sound synthesis practice and high efficiency [6].

Modern life is hard to believe without the use of nanotechnology. As mentioned before, the properties of nano-sized materials have led to their widespread use in various industries (Figure 8.1). Some of the major areas of nanotechnology applications are the production of textiles [7], industrial applications [8], agriculture and food science [9], electronics [10], environmental protection [11], renewable energy [12], biomedical [13], and healthcare science [14]. When faced with stronger but lighter and more durable materials [15], it will be straightforward to prioritize the use of nanomaterials. These properties are in addition to the extensive mechanical and

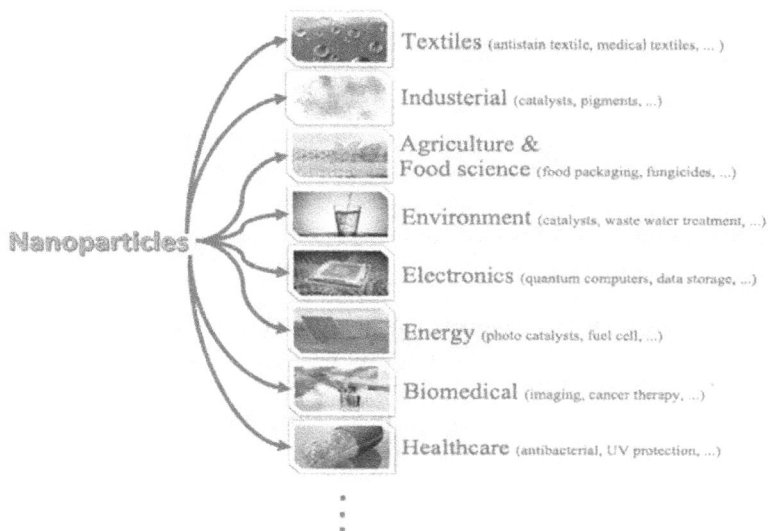

FIGURE 8.1   Various applications of nanoparticles.

electrical properties of nanomaterials, which have led to their popularity and widespread use in various industries.

8.2   ADVANTAGES OF GREEN SYNTHESIS OR BIOLOGICAL SYNTHESIS OF NANOMATERIALS

Generally, NP synthesis can be done by both chemical and physical methods. The physicochemical methods of nanomaterial synthesis include physical vapor deposition, sol-gel, radiofrequency plasma, pulsed laser, hydrothermal, combustion, etc. In addition to their advantages, these methods have many disadvantages and adverse effects, including the use of chemicals that endanger the environment and humans over time. For replacement, many methods have been proposed: the "green synthesis and/or biological synthesis" is one of the newest and least dangerous of these.

Green nanotechnology is NP synthesis by microorganisms – in other words, the use of microbial cells for the synthesis of nano-sized material [16]. In this non-hazardous method, NPs can be produced by phenols, terpenes, flavonoids, and proteins as secondary metabolites of plants and microorganisms [17–19]. The basic function of NP green synthesis is related to various reaction parameters, including pH, strain, temperature, substrate concentration, and solvent [20]. In previous studies, green synthesis of NPs was frequently reported from plants and plant-derived secondary metabolites; however, lately, microorganisms have also been proposed in NM synthesis [21]. Recent studies have shown that algae, *Actinomycetes*, bacteria, fungi, and yeast are used to synthesize nanomaterial for various aims [17, 20, 21]. Microorganisms utilize intra- and extracellular enzymes to synthesize NPs with different chemical compositions, sizes, shapes, and controlled monodispersity (of similar size and shape). It is an inexpensive, economically promising, and eco-friendly means for biosynthesis of NPs. Fungi have received much attention compared to other microorganisms. In other words, for some reason, fungi have the edge over plants and other microorganisms, including: (1) they can be mass-cultured in the laboratory; and (2) they can be grown in minimal medium and externally secreting enzymes [22].

8.3 MICROORGANISMS AS NATURAL FACTORIES FOR THE BIOSYNTHESIS OF NANOPARTICLES

The physical and chemical methods commonly used to make NPs generally consume much energy to provide the high pressure and temperature required to synthesize these materials. On the other hand, they use toxic substances, and residues may enter the environment during production. In particular, there is the potential for these contaminants to enter aquatic ecosystems such as drinking water systems, groundwater, and rivers [23]. Extensive application and consequently increasing demand for NPs as well as the need to use non-toxic, environmentally friendly, and biocompatible methods have made the use of microorganisms inevitable. In contrast to physiochemical processes, biosynthesized NPs are free from hazardous toxic chemical waste, which is essentially a desirable feature for biomedical applications. Another advantage of the biological synthesis of nanomaterials is that there is no need for additional processes to produce pharmacologically stable and active NPs. In addition, the biosynthesis method of nanomaterials is a shorter and faster path than their physicochemical production [24]. The formation of cadmium sulfide microcrystals in yeasts was one of the earliest reports of the biosynthesis of metallic NPs [25], and, thus far, the synthesis of NPs has been studied in numerous biological agents, including *Actinomycetes* and other bacteria, fungi, and plants.

Compared to bacteria, which need to be separated from a fluid extract to obtain NPs, fungi can accumulate metal ions by physiochemical and biological mechanisms, including extracellular binding to metabolites and polymers, binding to specific polypeptides, and metabolic accumulation [26]. Fungi are also significant sources of protein secretion compared to bacteria and *Actinomycetes*, which leads to higher efficiency in the production of NPs [26]. Therefore, due to these properties, fungi can be widely used for the rapid and safe biosynthesis of metal NPs. Accordingly, fungi are referred to as nano-factories for synthesizing metal NPs [27].

8.4 INTRODUCTION OF MYCONANOPARTICLES

The synthesis of NPs by fungi is called myconanotechnology, and the resulting NPs are called myconanoparticles, or myco-NPs. The term myconanotechnology was first coined by Rai and Bridge to describe the production of NPs by fungi and their application, especially in medicine. This science is a combination of mycology and nanotech, which has considerable potential for NP production due to the wide range of diversity in fungi [28]. Fungi have tremendous advantages over other microorganisms for the production of NPs. They are among the biological agents that have received most attention. Since fungi produce large amounts of protein and fungal proteins can hydrolyze metal ions, they can produce significant quantities of NP. In addition, fungi can be easily isolated and cultured. On the other hand, downstream processes and manipulation of fungal biomass are less complex than artificial methods. Short time, small size, and non-hazardous processes are critical parameters for accepting any type of technology for the production of NPs. Some fungi produce NPs with all these properties [27–30].

8.4.1 Biosynthesis of Fungi

8.4.1.1 Metallic Nanoparticles

Some fungi produce metabolites and enzymes that reduce metal ions to counteract the stress caused by exposure to metal ions. Such fungi are highly suitable candidates for the production of NPs. Exposing these fungi to metal salt solution and stimulation of osmotic stress conditions induce fungus to reduce metal ions and convert them to NPs [31]. The ability to be synthesized in the form of metal and metal oxide-based NPs has been observed in almost all metals. However, some metals, such as silver (Ag), gold (Au), aluminum (Al), cadmium (Cd), cobalt (Co), copper (Cu), iron (Fe), lead (Pb), and zinc (Zn), have frequently been used by researchers to synthesize NPs [32]. The production of silver and gold NPs has received more attention due to their particular importance and widespread use in various industries; consequently, it is clear that much research has been done on their biological production [33].

8.4.1.1.1 Silver Nanoparticles

The emergence of antibiotic-resistant and multidrug-resistant strains in microbial populations requires safe and alternative antimicrobials to overcome this problem, and NPs are promising in this regard [34]. Among different types of metal NPs, silver NPs have received much attention due to their unique physicochemical properties and broad-spectrum antimicrobial potential. The antimicrobial effects of silver NPs on human and plant pathogenic microorganisms have been proven [35]. Rai and Bridge suggested that silver NPs produced by fungi are a new generation of antimicrobials [28].

Aspergillus, with more than 350 species, is one of the fungi whose different species have been used repeatedly in producing silver NPs [30]. In a recent study, Chi et al. used metal-bearing *Aspergillus niger* biomass to produce AgNPs that showed anticancer potential [36]. Using marine-derived *A. brunneoviolaceus*, Mistry et al. produced AgNPs 0.72–15.21 nm in size, which showed strong antimicrobial properties against two Gram-positive and three Gram-negative bacteria. The biogenic NPs produced also showed antioxidant properties [37]. Othman et al. revealed that biosynthesized AgNPs using *A. terreus* had antimicrobial activity on PET/C fabrics. They also exhibited antitumor activity against different human tumor cell lines [38]. In a study conducted by Shahzad et al., spherical biosynthesized AgNPs using *A. fumigatus* exhibited antibacterial activity against multidrug-resistant bacterial strains, *Klebsiella pneumoniae, Acinetobacter, Pseudomonas aeruginosa,* and *Escherichia coli* [39]. Ottoni et al. isolated 20 different fungal strains from sugar beet farms and investigated their ability to biosynthesize silver NPs. Four strains, including *Aspergillus niger,* could convert silver ions to silver NPs, which showed antibacterial activity against *E. coli, Staphylococcus aureus,* and *P. aeruginosa* [40]. Overall, NPs synthesized by *Aspergillus* had generally regarded as safe (GRAS) status, which allows for their healthy use in the industry [30].

Fusarium is another fungus that has attracted the attention of researchers as a good candidate for the green synthesis of NPs. Different species of *Fusarium* have frequently been used for AgNP biosynthesis, and remarkable results have been obtained. Rodríguez-Serrano et al. used extracellular metabolites secreted by the soil fungal strain *Fusarium scirpi* to produce AgNPs. They evaluated the antimicrobial activity of these NPs and found that uropathogenic *E. coli* was inhibited from forming biofilms [41]. El-Sayed et al. used *F. solani* to produce AgNPs, copper, and zinc NPs. They assessed and demonstrated the biocidal activity of NPs against multidrug-resistant bacteria and mycotoxigenic fungi [42]. Ajah et al. used *F. graminaerum* to produce AgNPs, which had a significant antimicrobial effect on *Staphylococcus aureus, Pseudomonas aeruginosa, Streptococcus, Klebsiella, E. coli,* and *Candida albicans* [43]. In a study conducted by Ghabooli et al., it was shown that different *F. oxysporum* isolates had distinct capabilities in producing AgNPs. Notably, the NPs produced in the superior isolate maintained their stability even after 5 months [44]. Khalil et al. used *F. chlamydosporum* and *Penicillium chrysogenum,* biosynthesized AgNPs with great antifungal and antimycotoxin potential [45]. In a review article, Rai et al. discussed the mechanism and application of *Fusarium* AgNPs. They declared *F. oxysporum* as a fungus with a high potential for synthesizing AgNPs [46].

In addition to *Fusarium* and *Aspergillus,* there are numerous reports of the successful use of other fungi to synthesize AgNPs. Using *Penicillium verrucosum,* Yassin et al. produced polydisperse AgNPs with antifungal activity [47]. Naveen et al. utilized endophytic fungus, *P. radiatolobatum,* for green synthesis of AgNPs with the potential of antibacterial activity and cytotoxicity towards the human lung carcinoma cell line [48]. *P. citreonigrum, P. aculeatum,* and *P. italicum* are among other species of *Penicillium* which produce biogenic AgNPs with antimicrobial activity [27, 49, 50]. Many other fungi have been used to produce AgNPs, including *Cladosporium halotolerans, Colletotrichum, Botrytis cinerea, Trichoderma harzianum, T. gamsii, Rhizopus arrhizus, Sclerotinia sclerotiorum, Alternaria,* and *Arthroderma fulvum* [17, 40, 51–55].

In recent years, entomopathogenic fungi have attracted much attention in producing silver NPs. *Beauveria bassiana, Metarhizium anisopliae, Isaria fumosorosea,* and *T. harzianum* are entomopathogenic fungi with potential for the biosynthesis of silver NPs [56, 57]. Mushrooms are another group of fungi harboring the capability of NP biosynthesis. *Pleurotus* spp., *Ganoderma enigmaticum, Trametes ljubarskyi,* and *Bjerkandera* sp. [58–60] have been used as myconano-factories for the biosynthesis of NPs.

8.4.1.1.2   Gold Nanoparticles

The neutral nature of gold NPs and consequently their non-toxicity to humans have attracted the attention of many researchers. Due to this significant advantage, the use of AuNPs in biomedicine has received much attention; they have been widely used in biosensors and antimicrobial materials [33]. Ameen et al. [61] used the marine fungus *Alternaria chlamydospora* to produce AuNPs with antioxidant, antibacterial, and cytotoxicity activity. Clarance et al. utilized an endophytic strain of *Fusarium solani* for the biogenic synthesis of AuNPs. Biosynthesized AuNPs showed anticancer activity and had cytotoxicity on cervical cancer cells (HeLa) and human breast cancer cells (MCF-7) [62]. In addition, the anticancer activity of AuNPs, synthesized by endophytic fungus *Cladosporium* sp. against breast cancer cell line MCF-7, was observed on mouse models under *in vitro* and *in vivo* conditions [63]. Naimi-Shamel et al. produced gold nanoparticles (GNPs) using *Fusarium oxysporum*. Biosynthesized GNPs, which were then conjugated with tetracycline, had antibacterial activity against multidrug-resistant bacteria [64]. Domany et al. produced myco-AuNPs using *Pleurotus ostreatus* extracellular filtrate. Compared to commercial AuNPs, biosynthesized AuNPs had more antiproliferative properties and significantly reduced viable cells in the human liver cancer cell line (HepG2) and human colon cancer cell line (HCT-116) [25]. Using 29 thermophilic fungi, Molnár et al. compared three different approaches, the extracellular fraction, the autolysate of the fungi, and the intracellular fraction, for mycofabrication of AuNPs. Their results showed that the formation and size of AuNPs depend on the fungal strain and experimental conditions [65]. *Cladosporium oxysporum*, *Phoma* sp., *Chaetomium globosum,* and *Macrophomina phaseolina* have been used to produce biogenic AuNPs [29, 66–68]. There are also many other reports of the benefit and efficacy of fungi, which some researchers have listed in recent review articles [69, 70].

8.4.1.1.3 Other Metallic and Metallic Oxide Nanoparticles
Biosynthesis of other metallic and metallic oxide NPs has been performed by various fungal species, and research in this field is rapidly expanding. Table 8.1 presents the recent research on the production of NPs of copper, zirconium, platinum, copper oxide, zinc oxide, titanium dioxide, aluminum oxide, cobalt oxide, and cadmium sulfide, along with the general characteristics of NPs and the fungal species that produce them.

8.4.1.2 Non-Metallic Nanoparticles
In recent years, the use of fungi in synthesizing non-metallic NPs has also attracted the attention of researchers. Among them, selenium and tellurium have received much attention due to their unique photoconductivity properties and thermal conductivity. They are referred to as "E-tech" elements and are used to construct photovoltaic solar panels, semiconductors, and electronic devices [71]. Silicon, germanium, and tellurium are other non-metals whose NP production has been considered by fungi [72–75]. However, these elements are actually metalloid and have intermediate properties.

Many fungal species have shown the ability to transform non-metallic salts into NPs. Using culture filtrate, cell lysate, and crude cell wall of different *Trichoderma* spp., Nandini et al. biosynthesized selenium NPs (SeNPs) from 25 mM sodium selenite. These NPs showed antifungal properties against *Sclerospora graminicola,* and interestingly, by reducing the size of the produced NPs, the amount of antifungal properties increased [76]. *Trichoderma atroviride* was also used by Joshi et al. to synthesize mycogenic SeNPs with antifungal activity [77]. There are numerous reports

TABLE 8.1
Synthesis, Particle Size, and Application of Myconanoparticles

NPs	Fungi	Nanoparticle size (range and/or average) (nm)	Morphology	Application	Ref
Ag	Aspergillus niger	20.15–22.21 (21.38)	Spherical	Cytotoxic	[95]
	Aspergillus brunneoviolaceus	0.72–15.21	Spherical	Antibacterial	[96]
	Aspergillus terreus	13–49	Spherical	Cytotoxic	[97]
	Aspergillus fumigatus	0.681	Cube	Antibacterial Antifungal	[39]
	Aspergillus flavus	18.2	Spherical	Antibacterial	[98]
	Aspergillus niger	3.45–28.21 (10.31)	—	Antiamebic	[99]
	Aspergillus terreus	~13.80–2.0	Spherical	Antibacterial	[100]
	Aspergillus oryzae	Different sizes	Spherical	Antibacterial	[88]
	Fusarium scirpi	2–20	Quasi-spherical	Antibacterial	[101]
	Fusarium solani	7.65–18.89 (13.70)	Spherical	Antibacterial	[102]
	Fusarium graminearum	94	Spherical	Antibacterial Antifungal	[103]
Ag	Fusarium chlamydosporum	6–26	Spherical	Antifungal Cytotoxic	[104]
	Penicillium chrysogenum	9–17.5	Spherical	Antifungal	[104]
	Penicillium verrucosum	10–12	Irregular morphology	Antifungal	[105]
	Penicillium radiatolobatum	5.09–24.85	Spherical hexagonal	Antibacterial Cytotoxic	[106]
	Penicillium aculeatum	4–55	Spherical or approximately spherical	Antimicrobial Cytotoxic	[107]

Metal	Fungi	Size (nm)	Shape	Application	References
	Penicillium citreonigrum, Scopulariopsis brumptii	6–26	Spherical	Antibacterial	[108]
	Penicillium italicum	33	–	Antimicrobial	[109]
	Penicillium oxalicum	10–40	Spherical	Antibacterial	[87]
	Botrytis cinerea	5.1–13.95 (8.55)	Spherical	–	[53]
	Cladosporium halotolerans	20	Spherical	Antifungal, Antioxidant, Cytotoxic	[51]
	Colletotrichum	20–50	–	Antibacterial	[110]
Ag	Trichoderma harzianum	182.5 ± 6.9	–	Antifungal, Cytotoxic, Genotoxic	[35]
	Trichoderma gamsii, Rhizopus arrhizus, Aspergillus niger	30–100	Round-shaped	Antibacterial	[40]
	Sclerotinia sclerotiorum	10–15	Spherical	Antibacterial	[54]
	Alternaria	4–30	Spherical	Antibacterial	[111]
	Arthroderma fulvum	15.5 ± 2.5	Spherical	Antifungal	[112]
	Metarhizium anisopliae	28–38	Rod-like	Mosquitocidal	[56]
	Ganoderma enigmaticum, Trametes ljubarskyi	5–40	Spherical	Cytotoxic	[48]
	Bjerkandera	10–30	Spherical	–	[113]
Au	Alternaria chlamydospora	12–15	Spherical	Antibacterial, Antioxidant, Cytotoxic	[61]
	Fusarium solani	40–45	Spindle	Cytotoxic	[114]

(continued)

TABLE 8.1 (Continued)
Synthesis, Particle Size, and Application of Myconanoparticles

NPs	Fungi	Nanoparticle size (range and/or average) (nm)	Morphology	Application	Ref
Au	Fusarium oxysporum	22–30	Spherical, hexagonal	Antibacterial	[115]
	Cladosporium	5–10	Spherical	Cytotoxic	[116]
	Cladosporium oxysporum	72.32 ± 21.80	Quasi-spherical	–	[117]
	Pleurotus ostreatus	10–30	Spherical	Antibacterial Cytotoxic	[118]
	Phoma	10–100	Spherical	Antibacterial Antifungal	[67]
	Chaetomium globosum	20	Different shapes	Cytotoxic	[119]
	Macrophomina phaseolina	14–16	Spherical	–	[120]
Ag-Cu	Aspergillus terreus	20–30	Spherical	Antibacterial Antioxidant Cytotoxic	[61]
Cu	Fusarium solani	9.97–19.49 (13.42)	Spherical	Antibacterial	[121]
	Aspergillus niger	500	–	Antibacterial Cytotoxic Antidiabetic	[122]
CuCO₃	Neurospora crassa	150	Spherical	–	[123]
	Pestalotiopsis sp.				
	Myrothecium gramineum	40	Spherical	–	[123]
ZnO	Fusarium solani		Spherical	Antifungal	[121]
	Fusarium oxysporum	18–25	Irregular morphology	Bioethanol production	[124]
	Phanerochaete chrysosporium	5–200	Hexagonal	Antifungal Antibacterial	[125]
	Aspergillus sp.	~80–100	Spherical	Antibacterial Dye degradation	[126]

	Fungal species	Size (nm)	Shape	Activity	Reference
	Cladosporium tenuissimum	57	Hexagonal	Antifungal, Antibacterial, Cytotoxic, Dye degradation	[127]
PEGylated ZnNPs	*Monascus purpureus*	33.6–78.1	Rounded	Antifungal	[128]
ZnO-TiO$_2$	*Metarhizium anisopliae*	9.50	Spherical	Insecticide	[129]
TiO$_2$	*Tricoderma citrinoviride*	10–400	Polymorphic	Antibacterial	[130]
TiO$_2$	*Saccharomyces cerevisiae*	6.7 ± 2.2	Spherical	Antibacterial, Antifungal	[131]
Aluminum oxide	*Colletotrichum* sp.	30	Spherical	Antibacterial, Antifungal	[132]
CdS	*Aspergillus niger*	5	Spherical	Antibacterial, Cytotoxic	[133]
CdSQDs	*Fusarium oxysporum* f. sp. *lycopersici*	4.14	Spherical	Antibacterial	[134]
Cobalt oxide	*Aspergillus nidulans*	20.29	Spherical	–	[135]
Cobalt ferrite	*Monascus purpureus*	3–15 (6.50 ± 0.08)	Spherical	Antibacterial, Antifungal, Cytotoxic	[121]
Zr	*Penicillium* species	<100	Spherical	Antibacterial	[136]
Pt	*Penicillium pinophilum*	2–25	Different shapes	Antibacterial, Antifungal Cytotoxic	[137]

(*continued*)

TABLE 8.1 (Continued)
Synthesis, Particle Size, and Application of Myconanoparticles

NPs	Fungi	Nanoparticle size (range and/or average) (nm)	Morphology	Application	Ref
Se	*Trichoderma* spp. *T. asperellum* *T. harzianum* *T. atroviride* *T. virens* *T. longibrachiatum* *T. brevicompactum*	49.5–312.5	–	Antifungal	[76]
Se	*Trichoderma atroviride*	60.48–123.16	Spherical	Antifungal	[77]
	Trichoderma harzianum	60	Irregular	Antifungal Cytotoxic	[84]
	Monascus purpureus	30–70 (46.58 ± 1.54)	–	Antifungal Antibacterial Antioxidant Cytotoxic	[121]
	Monascus purpureus	48 ± 4.5	Spherical	Antibacterial	[81]
	Penicillium citrinum	Different Exposed to different doses of gamma radiation	Spherical	–	[51]
	Mariannaea	45.19 (intracellular SeNPs) 212.65 (extracellular SeNPs)	Spherical	–	[138]

Element	Species	Size	Shape	Activity	Reference
Se	*Ganoderma lucidum*	20–50	Spherical	–	[83]
	Lentinus edodes	50–320	Spherical	–	[83]
	Pleurotus ostreatus				
	Grifila frondosa				
	Aspergillus flavus and *Candida albicans*	51.5 and 64	Spherical	Antifungal	[139]
	Saccharomyces cerevisiae	4–51	Spherical	Antifungal	[20]
	Saccharomyces cerevisiae	4–7	Spherical	–	[140]
Se-Te	*Phanerochaete chrysosporium*	50–600	Needle-like, spherical	–	[72]
Te	*Phanerochaete chrysosporium*	20–465	Needle-like	–	[72]
	Aspergillus welwitschiae	60.80	Oval to spherical	Antibacterial	[141]
Ge	*Lentinus edodes*	50–250	Spherical	–	[83]
	Pleurotus ostreatus				
	Ganoderma lucidum				
	Grifila frondose				
Si	*Fusarium culmorum*	~40–~70	Spherical	–	[142]
	Aspergillus parasiticus	3–400	Pyramid	–	[143]
			Cubic		
			Spherical		

Note: NPs, nanoparticles; SeNPs, selenium nanoparticles.

on the successful use of *Saccharomyces cerevisiae* in producing SeNPs and their antimicrobial properties [20, 78]. Adding sodium selenate (Na$_2$SeO$_4$) to the culture filtrate of *Aspergillus flavus* and *Candida albicans*, Bafghi et al. produced SeNPs with antifungal activity [79]. Liang et al. used the fungi *Aureobasidium pullulans*, *Mortierella humilis*, *Trichoderma harzianum*, and *Phoma glomerata* to synthesize selenium- and tellurium-containing NPs. All fungi had the ability to reduce selenite and tellurite to elemental selenium and elemental tellurium, respectively. Selenium oxide and tellurium oxide were also biosynthesized by *T. harzianum* [71].

In addition to the above, various fungi such as *Monascus purpureus* [71, 80, 81], *Penicillium citrinum* [82], *Pleurotus ostreatus*, *Lentinus edodes*, *Grifola frondosa*, *Ganoderma lucidum*, *Agaricus bisporus* and *Agaricus arvensis* [83], *T. harzianum* [84], and *Mariannaea* sp. [85] have been promising in the production of SeNPs.

Vetchinkina et al. used *Lentinus edodes*, *Pleurotus ostreatus*, *Ganoderma lucidum*, and *Grifila frondose,* xylotrophic basidiomycetes, to produce selenium and germanium NPs [74]. Espinosa-Ortiz et al. utilized *Phanerochaete chrysosporium* to biosynthesize tellurium as well as selenium-tellurium NPs [72]. *Aspergillus welwitschiae* reduced potassium tellurite (K$_2$TeO$_3$) to TeNPs. The biogenic NPs showed antimicrobial activity against *E. coli* and *Staphylococcus aureus* [86].

8.5 INFLUENCING FACTORS ON MYCONANOPARTICLE SYNTHESIS

Although myco-NPs is an eco-friendly, cost-effective and simple procedure, it is necessary to optimize parameters that affect fungal growth as well as NP production to achieve stable, monodisperse, and biocompatible NPs. Various studies have shown that physicochemical characteristics of NPs are influenced by temperature, pH, the concentration of the metal precursor, amount of biomass, and culture medium [52].

8.5.1 TEMPERATURE

Temperature is one of the critical factors affecting NP synthesis by fungi. Although most studies have demonstrated that the rate of NP biosynthesis increases at a higher temperature, it should be noted that temperature also has particular importance in the size and stability of NPs; therefore, in finding the optimal temperature, this issue should not be overlooked [52]. Saxena et al. investigated the effect of temperature on AgNP mycofabrication by *Sclerotinia sclerotiorum*. After increasing the temperature from 20 to 80°C, the amount of NP biosynthesis increased; maximum synthesis was observed at 80°C [54]. The effect of the temperature (20–60°C) on the synthesis of AgNPs using extracts of *Penicillium oxalicum* was assessed by Rose et al. [87]. At higher temperatures, a narrow range of AgNP size was obtained, while at low temperature, different sizes of AgNPs were produced. Using *Aspergillus fumigatus*, Shahzad et al. obtained small-sized AgNPs at 25°C and increasing the temperature induced production of larger polydisperse NPs [39]. When the temperature was increased from 30 to 90°C, the rate of NP biosynthesis increased, and more uniform NPs were produced in less time. No biogenic NPs were produced at 10°C [88]. Azmath et al. showed that the biosynthesis of NPs by *Colletotrichum* was completed within

20 minutes at temperatures above 50°C [89]. At higher temperature, more small-sized NPs are produced, and the optimum temperature depends on fungal species. Apparently, at high temperatures, electrons can be transferred from free amino acids to silver ions. Nevertheless, at very high temperatures, capping agents may denature and decrease the stability of NPs [52].

8.5.2 METAL ION CONCENTRATION

Metal ion concentration is another crucial factor in the mycosynthesis of NPs and can affect the amount as well as the size and quality of NPs. Increasing the metal concentration from 0.5 to 1.5 mM doubled the bi-metallic Ag-Cu NP yield in *Aspergillus terreus* [90]. Rose et al. biosynthesized more AgNPs using *Penicillium oxalicum* by increasing $AgNO_3$ up to 1.5 mM. An increase beyond this concentration caused non-uniformity in NP size [87]. Osorio-Echavarría et al. [60] evaluated the concentration of $AgNO_3$ on the biosynthesis of AgNPs in *Bjerkandera*. They found that the most ion reduction, in the fungal extract and on the mycelium surface, occurred at 1 mM $AgNO_3$ concentration, 144 hours after incubation. The concentration of metal not only affects the quantity of produced NPs; it also regulates their size. Increasing the concentration of $AgNO_3$ from 1 to 8 mM reduced the size of AgNPs produced, while concentrations of 9 and 10 mM resulted in larger particles [88]. Ghabooli and Mirzaei showed that *Aspergillus flavus* produces more AgNPs at 2 mM $AgNO_3$ with an average size of 18.2 nm. These NPs were stable even 11 months after their biosynthesis and Fourier transform infrared demonstrated the presence of proteins as capping agents, leading to their stability [44]. Increasing the concentration of precursor salt to produce AuNPs using *Cladosporium oxysporum* raised the biogenic production. The optimum concentration was 1 mM but at concentrations of 0.5 mM, NP size was smaller [58].

Fungal extracellular proteins have an essential role in the reduction process. These functional groups enter into action after increasing the metal concentration and cause the production of smaller NPs. However, further increases in metal concentration due to a lack of functional groups can have the opposite effect and produce larger NPs [88].

8.5.3 pH

Since the biosynthesis of NPs is done through a reduction reaction by reducing agents, the presence of hydroxide ions as a factor that enhances this reduction capacity is significant [88]. This is why pH adjustment is so crucial in the biosynthetic capacity of NPs. Using pH 5–10, Rose et al. [87] showed that the mycofabrication of AgNPs by *Penicillium oxalicum* was very slow in an acidic range, and increasing pH resulted in smaller-sized AgNPs. At pH 7, more small-sized biogenic NPs were synthesized. In another study by Phanjom and Ahmed, AgNPs were not synthesized in pH range 4–5 in *Aspergillus oryzae* free-cell filtrate. More alkaline pH prompted smaller-sized AgNPs, and maximum synthesis was achieved at pH 10 in 30 minutes [88]. Sreedharan et al. investigated the effect of pH 5–9 on the biosynthesis of gold NPs by *Macrophomina phaseolina*. Increasing the pH from 5 to 8 led to more synthesis of gold NPs, but at pH 9 the mycosynthesis of NPs showed a negative trend and decreased [26]. In another study, the optimum pH for AgNP mycofabrication by *Aspergillus fumigatus* was 6 and increasing the pH resulted in larger-sized NPs [39]. Considering the extensive studies of researchers in this field, only a few have been mentioned here; it can be concluded that neutral to alkaline pH is most favorable for NP biosynthesis, but achieving the best pH depends entirely on the fungal species and even its strain.

8.5.4 FUNGAL BIOMASS

The amount of fungal biomass is another factor that has a significant effect on the properties of NPs produced by fungi. In some cases, an increase in fungal biomass has led to an increase in NP production, while in other cases, the opposite has been observed. *Cladosporium oxysporum* increases the concentration of fungal biomass, and consequently, the extracellular metabolites produce more AuNPs with smaller sizes [58].

Osorio-Echavarría et al. evaluated the effect of fungal growth time (3–8 days) on AgNP production by *Bjerkandera,* and the best results were obtained after using an 8-day-old culture [60]. Using 5–25 g biomass of *Penicillium oxalicum,* Rose et al. found that more biomass resulted in more small-sized and uniform AgNPs [87]. Saxena et al. showed that the increase in *Sclerotinia sclerotioum* biomass increased the reducing agents and enhanced the biosynthesis of AgNPs [54]. On the other side, Shahzad et al. investigated the culture age (3–7 days) and biomass weight (1, 4, 7, 10 g) of *Aspergillus fumigatus* in the biosynthesis of AgNPs [39]. The use of 7-day-old culture had the best outputs, and the increase in biomass (up to 7 g) led to the production of more AgNPs of smaller sizes, which could be due to the secretion of more enzymes. Further increase in biomass increased the size of the NPs produced, which apparently occurred due to saturation of the environment.

8.5.5 CULTURE MEDIUM

The composition of the culture medium can significantly affect the growth of fungus and, consequently, the metabolites produced by them. Since the production of NPs is a biological process, the properties of the particles will also be affected by the culture medium.

Comparing different culture media, Ghabooli and Mirzaei found that potato dextrose broth (PDB) was more suitable for producing AgNPs by *Aspergillus flavus*, due to the production of more biomass and the effect of the medium on the quantity and composition of secreted proteins [44]. To investigate the impact of media on AgNP biosynthesis, Saxena et al. cultured *Sclerotinia sclerotiorum* in different media [54]. PDB enhanced the production of NPs, probably because of more reducing agents. Figure 8.2 depicts effective factors in the mycosynthesis of NPs.

8.6 MECHANISM OF MYCOSYNTHESIS OF NANOPARTICLES

There are two basic methods for making NPs: the top-down and the bottom-up methods. The formation of a structure through the arrangement of atoms or molecules

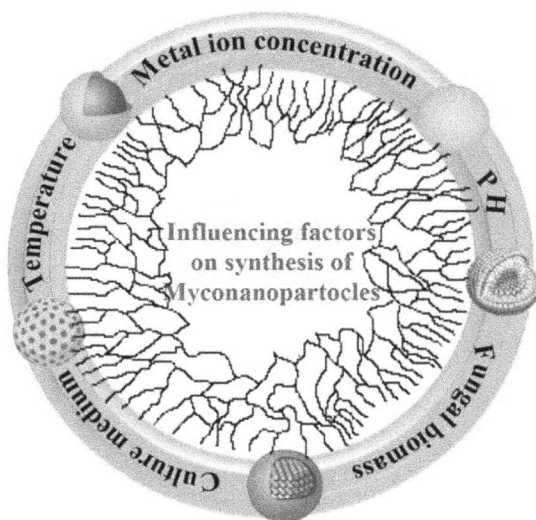

FIGURE 8.2 Effective factors in the mycosynthesis of nanoparticles.

next to each other involves the concept of bottom-up approaches. In the top-down approach, NPs are produced by reducing the size of the primary bulk material by applying physical or chemical methods. Fungal synthesis of NPs, which is classified in the bottom-up approach, occurs both intracellularly and extracellularly [91]. The exact mechanism of NP biosynthesis is not yet known, but there is no doubt that the enzymes and proteins secreted by the fungi play a significant role in this regard, and the primary mechanism behind this process is the reduction reaction [33]. Different microorganisms use diverse methods to synthesize NPs, but in all cases, there are three main steps: nucleation, growth, and stability of NPs. The fungus first traps the metal and builds the nucleus by a reduction reaction, and then the nucleus grows and stabilizes with the coating agents. These ligands prevent further growth and coagulation of particles. Sometimes a chemical compound acts as both a reducing agent and a coating agent; citrates and amino acids are among these compounds [65]. In the intracellular production of NPs, metal ions primarily are necessary to be transported into the cells, and this occurs through electrostatic attraction between positively charged metal ions and negatively charged cell walls. After transportation of the ions into the cell, enzymatic bio-reduction converts metal ions to NPs; fungal proteins then act as coatings and stabilize the resulting NPs [92]. The synthesis of NPs in the intracellular system results in smaller NPs than in the extracellular system, but the fact is that extracting intracellular NPs is more complex and requires more cost [70]. In extracellular production, the fungus secretes the required enzymes into the environment or on the cell wall, and the metal ions in the environment are reduced by these secreted enzymes [21]. NADPH-dependent nitrate reductase is a key enzyme in the reduction of metal ions. In the presence of NADPH, nitrate is converted to nitrite by nitrate reductase. The electron released from NADPH can reduce Ag^+ to Ag^0, leading to the production of NPs [93]. Various studies have investigated the role of quinones in the

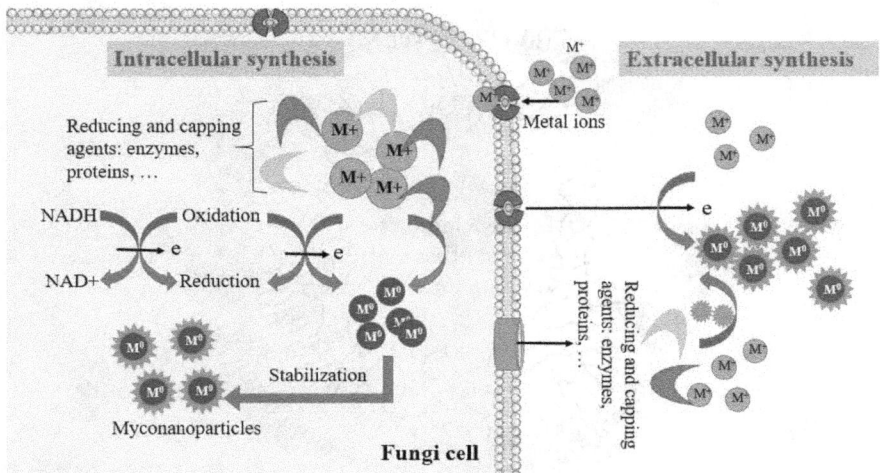

FIGURE 8.3 Schematic illustration of intracellular and extracellular mycosynthesis of nanoparticles.

biosynthesis of NPs. Both NADPH and quinone were initially thought to provide the required electrons, but later studies showed that the NP production process required NADPH, and that hydroxyquinone probably acted as an electron shuttle, transferring the electron generated during the reduction [46]. Many studies have also investigated the role of other proteins and enzymes in the biosynthesis of NPs by fungi, but among all the hypothetical mechanisms available, the nitrate reductase mechanism is the most accepted [46]. Nevertheless, Hietzschold et al. showed that NADPH alone acts as a reducing agent for silver nitrate to produce silver NPs. The enzyme-free reaction was more efficient, and the particles were smaller, monodispersed, and more stable [94]. The general mechanism of NP synthesis is illustrated schematically in Figure 8.3.

NP biosynthesis and their characteristics depend on the interaction of fungal species, type of metal ions, and environmental conditions. This has led to many ambiguities about the exact mechanism of NP biosynthesis by fungi, which requires further investigation [94].

8.7   APPLICATION OF MYCONANOPARTICLES

Myco-NPs have various potential applications in agriculture, health, and pest control. Synthesis based on fungi may be advantageous in production due to the large quantities of metabolites produced. Another factor to consider is the capacity of fungi to produce antibiotics that could be contained in the capping and act in synergy with the NP core [35].

8.7.1   MEDICINAL APPLICATIONS

Many studies of biogenic synthesis of NPs using fungi have shown promising results for the application of these systems in controlling pathogenic fungi and bacteria, combating cancer cells and viruses, and providing larvicidal and insecticidal activities. With the production of reactive oxygen species by NPs, bacterial growth is directly inhibited; it contacts the cell wall and causes progressive metabolic responses [144–148]. One of the factors determining NP antimicrobial potential is their size. Smaller particles are much more effective [123]. The smaller the size, the better the penetration, so NPs can penetrate the bacterial cell membrane, damage the respiratory chain, alter permeability, cause DNA and RNA damage, affect cell division, and lead to cell death [149, 150]. Also, myco-NPs interact with the thiol groups on essential enzymes, releasing ions to form complexes with nucleotides, damaging the DNA of microorganisms and inhibiting the activity of DNase [123, 151]. Myco-NPs can be an inexpensive and safe option for treating systemic and surface fungal infections, enabling the control of resistant fungus [152]. The large surface area of myco-NPs and the release of ions can contribute to high antimicrobial activity. The toxic ions bind to protein-containing sulfur, affecting cell permeability and leading to an alteration in the DNA replication process.

The inactivation of some enzymes is also caused by the binding of NPs with thiol groups. The oxidative stress resulting from this inactivation affects electron transfer and protein oxidation [30, 153, 154]. Control of *Staphylococcus aureus* and *Klebsiella pneumoniae* was performed *in vitro* using AgNPs synthesized with *Trichoderma harzianum* [155]. The inhibition rates were concentration-dependent, with the Gram-negative organisms (*K. pneumoniae*) showing higher sensitivity. The potential of myco-NPs using *Guignardia mangiferae* to control Gram-negative bacteria, with effects including increased permeability, altered membrane transfer and release of nucleic acids [156]. Shrivastava and Dash reported that the lower effects of silver NPs on Gram-positive bacteria may be because the peptidoglycans that compose the cell wall act as a barrier that prevents the internalization of the NPs [157]. However, in some studies, the NPs exhibited inhibitory effects against this bacteria type. In evaluation of the antimicrobial activity of silver NPs synthesized using *Aspergillus niger*, Gade et al. observed inhibitory effects against the bacteria *E. coli* and *S. aureus* that were equivalent to those of the antibiotic gentamicin, with the Gram-positive bacterium (*S. aureus*) showing greater sensitivity [158]. AgNPs have been used in combination with antibiotics and antifungals, representing a possible solution to the problem of resistance toward these drugs used in health.

AgNPs were synthesized using *Candida albicans* and their effects were evaluated when used alone or in combination with the antibiotic ciprofloxacin against *Staphylococcus aureus, Escherichia coli, Bacillus cereus, Vibrio cholerae*, and *Proteus vulgaris*. The activity of the antibiotic increased when it was used together with the NPs, while the latter also showed antimicrobial potential when they were used alone [58]. The antimicrobial and antifungal activities of AgNPs synthesized using the filtrate from *Aspergillus flavus* were evaluated by Fatima et al. [154]. The NPs effectively controlled the bacteria *Bacillus cereus, B. subtilis, Enterobacter aerogenes, Escherichia coli*, and *Staphylococcus aureus*, with *B. subtilis* and *E. coli* being most sensitive. The activity was concentration-dependent, with better results achieved for the NPs in combination with the antibiotic tetracycline, rather than on their own. Concentration-dependent activity of the NPs was also observed against the fungi *Aspergillus niger* and *Trichoderma harzianum*.

8.7.1.1   Drug Delivery Systems

Drug delivery to the target site is one of the critical applications of nanotech. Drug delivery systems in agriculture can be nanotech with the combination of the following characteristics: spatially targeted, preprogrammed, time-controlled, self-regulated, remotely regulated, or multifunctional characteristics to avoid biological barriers to successful targeting [159]. Drug delivery systems can also have the capability to examine the effects of the delivery on plants, soils, the environment, insecticides, fungicides, and insects.

The drug delivery system has massive potential for improving agro-chemical efficiency in agriculture. The drug delivery system has a huge potential for improving the efficiency of fungicides in agriculture systems. This is a new technology that improves the protection of plants against pests and diseases [159]. Nanofungicides, nanopesticides, and nanoherbicides are being used extensively in agriculture [160]. The formation of microorganism-based pesticides, fungicides, and herbicides could be the replacement, at least partly, for chemical fertilizers. Remote activation and monitoring of intelligent delivery systems can assist agricultural growers of the future in minimizing fungicide and pesticide use. Records based on microorganisms for developing agricultural chemicals, pesticides, fungicides, plants, soil, and seed treatment are available [161]. There is also evidence that investigation for new formulations of plant protection products with quantitatively high application potential is continuing.

Therefore, using nanomaterials based on microorganisms will be preferable to chemical-based nanomaterials due to their low cost, low risk, and faster preparation.

8.7.2   Agriculture Application of Myconanoparticles

Changing weather patterns, urbanization, and environmental concerns, including environmental contamination and pesticide accumulation, contribute to an ever-increasing demand to supply healthy and enough food for the growing global population because natural resources such as arable lands, water, and fertile soils are reducing as a result of overpopulation [162]. Crop productivity is affected by calamitous plant insects and diseases that reduce crop quality and quantity, exacerbating the food supply gap and causing millions of people to go hungry. Overall, in phytopathology, the term "pesticide" covers insecticide, herbicide, rodenticide, fungicide, plant growth regulator, and nematicide, so the term "pest" includes any weed, insect, or pathogenic disease that is damaging for plants. Typically, the use of conventional pesticides results in excessive residual toxicity and environmental concerns because pesticides are frequently used without regard for preventive measures and dangerous aspects. Indeed, in addition to the cost of chemical inputs, adverse side effects on human health, biodiversity, and the environment can lead to even higher costs in the future to manage pests [142]. Since conventional pesticides have non-target delivery, they are harmful for non-targeted specific species, valuable insects, and antagonistic microorganisms. There is a need to apply greener and safer technological approaches to overcome these constraints and achieve safer products with more qualifications and better flavor. Therefore, many plant pathologists are working to find a solution based on novel technologies to make such a desire come true. In this regard, applying nano-based approaches can be an efficient protocol to protect agriculture products from phytopathogens in a greener manner. In many fields such as agriculture, nano-technology has rapidly become a prominent study topic and it has great potential to develop efficient, sustainable, eco-friendly, and cost-effective approaches. Due to their high surface-to-volume ratio and slow-release, targeted delivery, nanomaterials offer more efficient, safer, and greener inputs such as nanopesticides, nanoherbicides, and nanofertilizers [163]. Numerous documents have established the efficacy of different NPs against pests. However, the effect of NPs in alleviating phytopathogens is associated with shape, size, the dose of NPs, duration of exposure, and plant species exposed to them. The above feature leads to higher performance of nano-based inputs than traditional pesticides. In the seed germination, seed priming, seed germination, and growth stimulation processes, many NPs show promise as potential novel treatments [164]. In addition, in agriculture-based nanotechnology, the applicability of agrochemicals under the smart delivery system, controlled release of pesticides, encapsulation capability of agricultural inputs, and biocompatibility and biodegradability improve the efficiency of agrochemicals in agricultural systems [165]. In green nanotechnology, bioactive molecules originating from plants, bacteria, yeast, and fungi can act as regenerative agents for the biosynthesis of NPs [166]. The potential of fungi to grow on commonly available and affordable substrates in the laboratory and their ability to produce a wide variety of metabolites such as enzymes have generated substantial interest in them as qualified candidates for study to synthesize NPs [167, 168]. The antipathogenic properties of myco-NPs provide their potential for pest management. These nano-molecules are not only limited to plant protection; they have a wide array of applications in other aspects. Myco-NPs have been reported as promising functional elements in the remediation of toxic chemicals from environmental resources [169]. MycoNPs as micronutrients can be a preferable alternative to conventional fertilizers. They can help enhance plant growth because these nutrients are taken up more effectively by the plant; as a result, myco-NPs can help reduce the toxicity of the soil created by chemical fertilizers [162]. Further, it is expected that the capacity of myco-NPs in targeted delivery of biomolecules or as a diagnostic tool and nano-sensors for disease detection position them as the most critical and potent tool as an alternative in a pesticide-saturated market and minimize unsafe agriculture output [77, 170]. The major application of myco-NPs in plant protection is depicted in Figure 8.4.

The myco-NPs as micronutrients can be a preferable alternative to chemical fertilizers. They can help enhance plant growth because these nutrients are taken up more effectively by the plant; as a result, myco-NPs can help reduce the toxicity of the soil caused by chemical fertilizers.

8.7.2.1 Role in Management of Pests

As plants grow, they are attacked by different pests, including insects, fungi, bacteria, nematodes, and viruses during the different stages of their life cycles. Stresses induced by these ruinous biotic agents lead to reduced photosynthetic capacity, slowed growth, and a decreased quality and quantity of crops. Severe attacks by pests cause severe injury to the host plant, resulting in plant death [171–173]. Ecological changes in the environment result in the emergence of new pathogenic

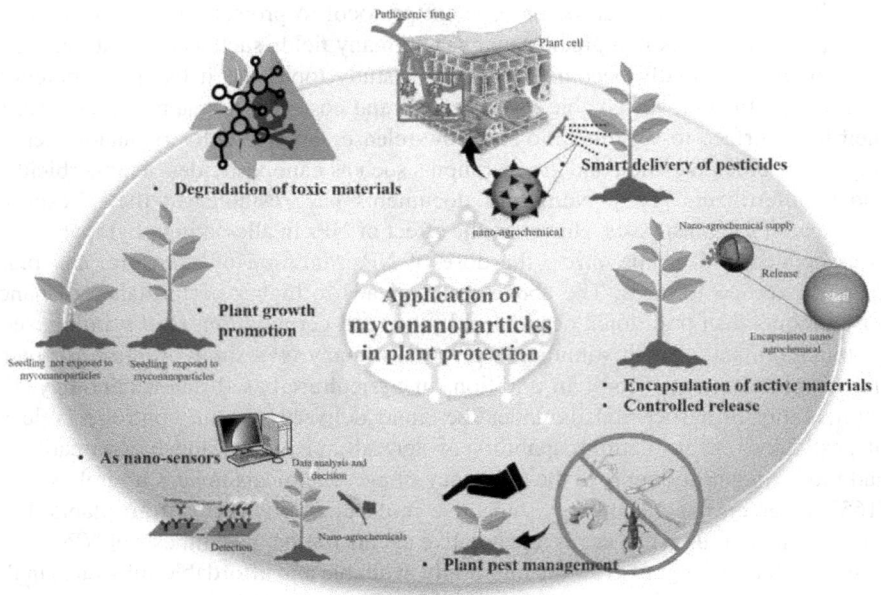

FIGURE 8.4   The major application of myconanoparticles in plant protection.

strains [159]. Protecting agricultural crop species against new strains of invasive phytopathogens is essential. On the other hand, the indiscriminate use of pesticides and the growing trend of pest resistance to pesticides have attracted investigators' attention towards more effective and greener alternative materials to manage crop pests [174]. In recent decades, rapid developments in nanopesticide research have inspired some scientific communications to consider nanotechnology as a field of rapid advancements to find a way to solve possible concerns in crop protection issues. Organic nanomaterials (such as carbon-based nanomaterials), inorganic nanomaterials (such as silver, gold, copper, titanium dioxide, zinc oxide, alumina, and silica nanomaterials), and nanocomposites (such as nanopolymers and chitosan-based nanomaterials) are three types of nanopesticides [175]. There has been evidence that these nanomaterials can act as insecticidal, bactericidal, fungicidal, nematicidal, and antiviral compounds against an extensive range of pests [176, 177]. Additionally, smart delivery of green synthesized nanomaterials to a target site in the host plant at appropriate concentrations results in the plant's mitigation response to plant pests [164]. In this regard, numerous research has been conducted on applying myco-NPs as pesticides to suppress calamitous plant pests. Different myco-NPs have been proven to be effective pesticides. Extensive research has hinted at the biosynthesis of most nanomaterials such as silver [178], gold [179], quantum dots [180], selenium [77], silica [83], titanium [181], platinum [182], magnetite [102], and cobalt [42] using fungi. Some of these NPs were successful at suppressing plant pests. Many of these NPs have been shown to have antipathogenic medicinal properties, making them a promising alternative for controlling plant pests.

8.7.2.1.1 Myconanoparticles as Fungicides and Bactericides

Bacteria and fungi phytopathogens provide disease epidemics, resulting in enormous financial losses and grave implications for the secure supply of a region's food. Traditional management approaches limit the effective control of plant pathogenic bacteria and fungi and provide prospects for their expansion due to resistance to conventional bactericides and fungicides. Therefore, developing novel techniques to combat pathogens becomes critical when disease management fails [183]. Nanoscience and nanotechnology have emerged in the recent decade, allowing researchers to investigate the bactericidal properties of nanomaterials against bacteria and fungal phytopathogens. Most research has investigated the bactericidal effects of myco-NPs against medicinal bacteria [48, 184–186]. There are few reports about myco-NPs' antibacterial activity against bacteria phytopathogens.

Gold NPs (AuNPs) are a perfect option for various applications, including agriculture and medicine, due to their stability, functionalization, low toxicity, and ease of detection [187]. In a study, *Phoma* sp. was evaluated to check its ability to biosynthesize AuNPs. In this study, Soltani Nejad et al. [179] proved the antibacterial and antifungal effect of extracellularly biosynthesized AuNPs against rice pathogens, *Rhizoctonia solani* AG1-IA, and *Xanthomonas oryza*e pv. *oryzae*. Their findings showed that AuNPs could inhibit the formation of sclerotia of *R. solani* up to 93% at 80 µg/mg concentrations. Numerous investigations indicated that silver NPs (AgNPs) demonstrated substantial antipathogenic activity, especially against plant pathogenic bacteria and fungi [120, 188]. According to research, silver's antibacterial and antifungal properties can be attributed to these nanomaterials inducing oxidative stress, protein malfunction, and DNA and membrane damage [116]. In the most recent study in 2022, Zaki et al. [189] reported the ameliorative role of mycogenic AgNPs in relieving damping-off pathogens of cotton, namely *Fusarium fujikuroi*, *Macrophomina phaseolina*, and *Rhizoctonia solani in vitro* and *in vivo*. At 40 and 100 µg/mL, AgNPs dramatically inhibited the radial expansion of colonies in all three pathogens in an *in vitro* test. In addition, under greenhouse pot conditions, the survival percentage of cotton seeds coated with AgNPs cultivated on contaminated soil was higher than the negative control that was not coated with AgNPs. These results were competitive with the treatments that received chemical fungicide treatment, and clearly prove that mycogenic AgNPs can be a practical alternative for chemical fungicides suppressing phytopathogens (Figure 8.5).

SeNPs are biologically safe and biocompatible [190]. SeNPs are an attractive and practical choice due to various characteristics, including antibacterial, antifungal, and anticancer activities [191–193]. Se is also a crucial element for human health [194]. Joshi et al. [77] assayed the antifungal activity of mycogenic SeNPs mediated by *Trichoderma atroviride* against three phytopathogens. In Petri plate conditions, they investigated biosynthesized SeNPs against *Pyricularia grisea* at 50, 100, and 200 ppm concentrations. SeNPs reduced the radial growth of the colony of *P. grisea* at 100 and 200 ppm. Also, they evaluated the antifungal activity of SeNPs by leaflet assessment at 10, 25, 50, and 100 ppm concentrations against *Colletotrichum capsisi* and *Alternaria solani* on tomato and chilli leaves. Mycogenic SeNPs effectively suppressed the growth expansion of *C. capsisi* and *A. solani* on leaves at 50 and 100 ppm (Figure 8.6).

FIGURE 8.5 Antifungal activity of mycogenic silver nanoparticles (AgNPs). (a) *In vitro* antifungal effect of selenium nanoparticles against three plant pathogens and significantly reduced growth of the fungal colony at 40 and 100 µg/mL concentrations; (b) *in vivo* assessment of AgNPs; as a negative control, uncoated cotton seeds were sown in sterilized soil infested by three fungal pathogens, as a positive control, uncoated seeds were sown in sterilized soil infested with Maxim XL and Moncut fungicides, and seeds coated with AgNPs were sown in infested soil [189].

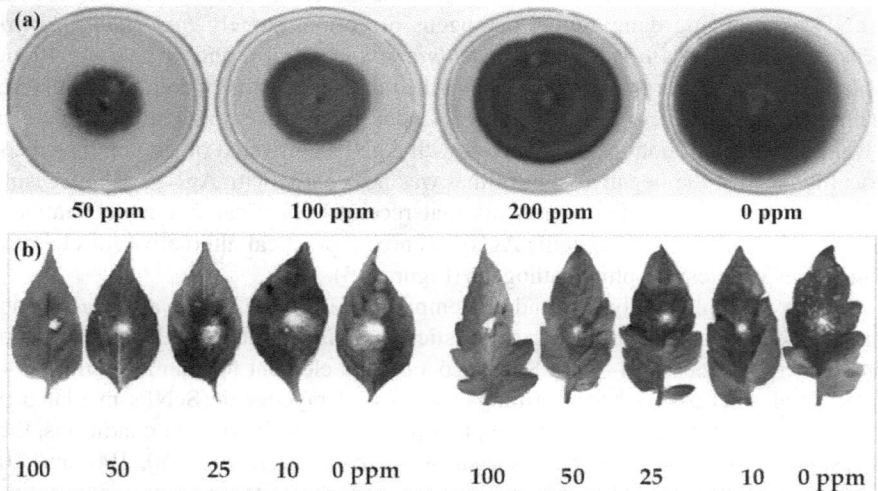

FIGURE 8.6 Antifungal effects of selenium nanoparticles (SeNPs) biosynthesized using *Trichoderma atroviride*. (a) *In vitro* effective antifungal activity of SeNPs against *Pyricularia grisea* at 100 and 200 ppm; (b) *in vitro* leaflet antifungal activity of SeNPs against *Colletotrichum capsisi* and *Alternaria solani* on chilli (left) and tomato (right) leaves at four doses. Mycogenic SeNPs effectively suppressed the growth expansion of *C. capsisi* and *A. solani* on leaves at 50 and 100 ppm [77].

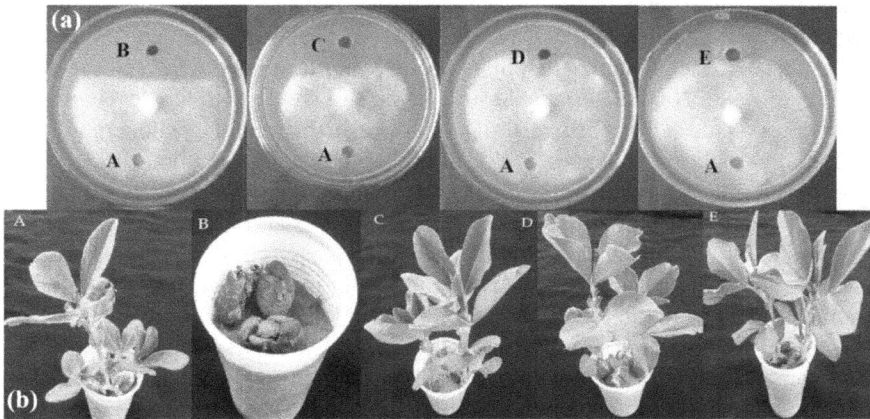

FIGURE 8.7 The anti-oomycete activity of zinc nanoparticles (Zn-NPs) synthesized using *Trichoderma longibrachiatum*. (a) *In vitro* anti-oomycete activity of Zn-NPs; A, B, C, D, and E wells containing distilled water, metalaxyl, ZnNPs at 20 ppm, ZnNPs at 15 ppm, ZnNPs at 10 ppm concentrations, respectively; (b) anti-oomycete activity of ZnNPs in the pot; (A) *Vicia faba* seed sown in sterilized soil (as control); (B) seeds sown in soil infested with *Pythium aphanidermatum*; (C–E) seed sown in soil infested with *P. aphanidermatum* and irrigated with NPs (10, 15, and 20 ppm), respectively, after 4 weeks [196].

In a recent assessment, in 2021, the anti-oomycete activity of SeNPs biosynthesized using *T. atroviride* was proved against *Phytophthora infestans*, the causal agent of late blight in tomato and potato. Seed priming with SeNPs elicited resistance against late blight in tomato seedlings [195]. The same research found that SeNPs made from six species of *Trichoderma* spp. helped control downy mildew disease in pear millet caused by *Sclerospora graminicola*. These myco-NPs effectively inhibited the maturation, sporulation, and zoospore viability of *S. graminicola* [76]. Numerous investigations indicated that AgNPs demonstrated substantial antipathogenic activity, especially against plant pathogenic bacteria and fungi [120, 188]. Moustafa and Taha [196] proved that mycogenic zinc NPs have excellent anti-oomycete potential *in vitro* and *in vivo*. Zn(II) complex in the nanoscale biosynthesized using *Trichoderma longibrachiatum* suppressed colony growth of *Pythium aphanidermatum* in 20 ppm concentrations *in vitro*. The colony growth rate was inversely related to the concentration tested (Figure 8.7a). It was found that by applying nano-Zn(II) complex at 20 ppm concentration by an irrigation method *in vivo*, the damping-off caused by *P. aphanidermatum* was effectively suppressed by 73.8% compared to the control pot that does not irrigate with NPs (Figure 8.7b).

8.7.2.1.2 Antipathogenic Mechanisms of Myconanoparticles

The mechanism of myco-NPs appears to be similar to that of other nanomaterials. However, the exact mechanism of action of nanomaterials against plant pests has not been fully grasped, although several theories have been proposed. The starting point for the function of nanomaterials in the face of plant pests begins with their attachment on the surface of the plasma membrane or cell wall of a pathogenic micro-organism [197]. Electrostatic attractions cause their attachment to the cell membrane or cell wall [198]. Nanomaterial adhesion on the surface leads to the development of perforations on the cell surface, resulting in disorganization of the cell membrane and leakage of cytoplasmic contents [199]. In prokaryotic cells, membrane breakup can disrupt bacterial cells' respiratory chain. According to one theory, in addition to the nanomaterials themselves, ions released from nanomaterials in aqueous environments can be toxic for live cells. Bapat et al. [200] claimed that the antibacterial mechanism of AgNPs could be related to the release of silver ions in aqueous media. Additionally, nanomaterials may disrupt the function of efflux pumps in the cell membrane. They bond to sulfur-containing proteins, preventing them from adequately performing in the membrane and interfering with permeability [201, 202].

In another theory, nanomaterials penetrate the microorganism cell and impair protein and nucleic acid synthesis pathways. They damage DNA or RNA and diverse organelles of the cell, interfering with the metabolic processes of cells and ultimately resulting in cell death [203]. Denaturation of the 30S subunit of ribosomes may alter the inactivation of enzymes in prokaryotic cells [204]. Pareek et al. [205] stated that Ag ions could significantly suppress protein synthesis by denaturing cytoplasmic ribosomal subunits. In addition to non-oxidative mechanisms, oxidative stress by generating reactive oxygen species (ROS) is an effective mechanism through which NMs cause cell damage. By reducing oxygen and producing diverse ROS molecules such as superoxide (O_2^-), hydrogen peroxide (H_2O_2), hydroxyl (OH^-), and hydroxyl radical ($\cdot OH$), nanomaterials can cause underlying damage to cell components such as protein, cell membrane, or nucleic acid. Damaging these components can eventually result in the cell's death [205, 206]. The antipathogenic mechanism of myco-NPs is depicted schematically in Figure 8.8.

8.7.2.1.3   Myconanoparticles as Insecticides

Toxicity to plants, soil, and environmental components is a consequence of the indiscriminate use of chemical agrochemicals. In addition, overusing synthetic chemical pesticides to manage pests results in insect resistance to pesticides. In this situation, nanopesticides are a supreme replacement for synthetic agrochemicals since they are greener and safer and can act in a target-specific way in comparison to traditional pesticides for the biological system. Nano-based pesticides hold tremendous promise for managing insect pests in current agriculture; however, it is imperative to evaluate their positive and negative impacts on pest insects carefully. Nanomaterial-based pesticides can be sprayed on to insect pests or crops to feed pests or enter insect bodies. They also can be used as insect repellent sprays. For instance, it has been stated that nanotubes loaded with alumino-silicate can adhere to plant surfaces. In contrast, other nanotube elements can attach to the pest's surface hair, eventually penetrating the body and causing damage [113, 207]. Myconanotechnology represents a novel pathway in synthesizing insecticidal nanomaterials and provides an alternative to pesticides for controlling plant pests. The efficiency of myco-NPs against insect pests has been demonstrated in studies, indicating that these particles have a critical role in the effective management of plant insect pests [162].

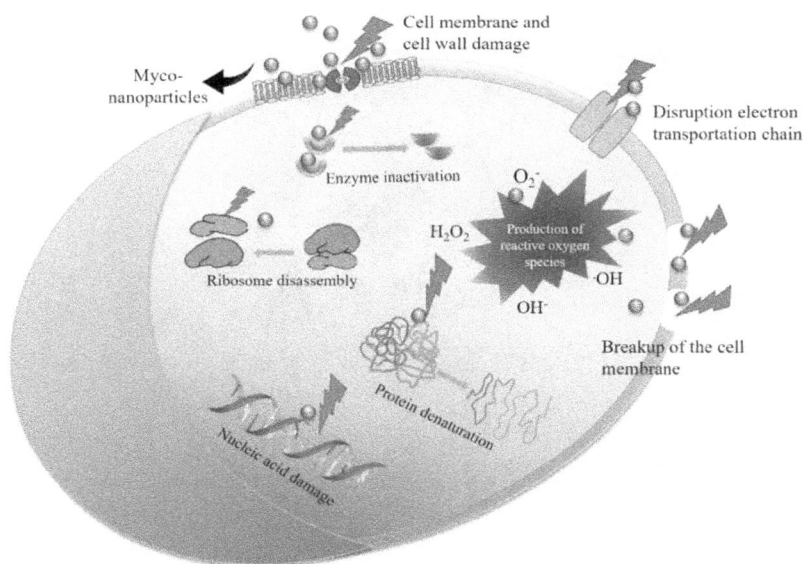

FIGURE 8.8 The antipathogenic schematic mechanism of myconanoparticles.

In addition to the excellent and proven antibacterial and antifungal activity of AgNPs, numerous research studies have dealt with the insecticidal effect of these NPs [208]. In this regard, the insecticidal activity of *Aspergillus niger*-derived mycogenic AgNPs was assessed by Shukla et al. [209]. They evaluated the larvicidal effect of AgNPs against first instar white grub (*Holotrichia* sp.) larvae under *in vitro* conditions. The results revealed the excellent biocidal effect of AgNPs, and LD_{50} was recorded at 9.03 ppm. There has been a recent uptick in interest in nanoscale SeNPs for plants because of their effectiveness in alleviating various biotic stressors and the capacity to control plant pests [106]. In a study by Amin et al. [210], *Penicillium chrysogenum* was applied as a reducing agent for the biosynthesis of SeNPs. Evaluation of the larvicidal activity of five concentrations (5, 10, 15, 20, and 25 ppm) of SeNPs against *Agrotis ipsilon* revealed that SeNPs at 25 ppm concentrations have the highest larvicidal impact on all five instars of larvae (first, second, thirrd, fourth, and fifth). In similar research, in 2021, insecticidal activity of SeNPs myco-synthesized using *Trichoderma* sp. was assessed against *Spodoptera litura* larvae. Larvae exposed to SeNPs were affected by significant mortality after 48 hours. LC_{50} and LC_{90} values were recorded at 39.73 and 142.83 ppm, respectively. These nanomaterials have a remarkable antifeedant effect at 100 µg/mL [211].

As regards other nanomaterials, various works hint at the insecticidal activity of titanium dioxide NPs and Zn-based NPs [212–214]. In a comparative study, the effect of mycogenic monometallic (TiO_2NPs and ZnONPs) and bimetallic NPs (TiO_2-ZnONPs) was assessed against third-instar *Spodoptera frugiperda* larvae at three concentrations (25, 50, and 100 µg/mL) [142]. Mycogenic monometallic and bimetallic NPs were synthesized using *Metarhizium anisopliae*. The published

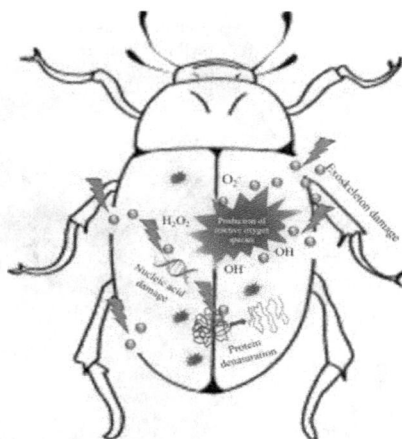

FIGURE 8.9   Schematic depiction of the insecticidal mechanism of myconanoparticles.

findings exhibited the highest mortality rate (85%) at a 100 g/mL dose after 48 hours of treatment with TiO_2-ZnONPs compared to monometallic NPs on corn plants. In addition, the TiO_2-ZnONPs significantly showed an antifeedant effect. The highest feeding activity belonged to fresh castor leaves treated with TiO_2-ZnONPs at a 25 μg/mL (43%) concentration. The lowest feeding activity belonged to leaves treated with 100 μg/mL.

8.7.2.1.4   Insecticidal Mechanism of Myconanoparticles

Similarly to other nanomaterials, it is possible to generalize the insecticidal mechanism of nanomaterials to myco-NPs. To estimate accurately the effectiveness of myconanomaterials as insecticides in the real world, it is critical to have detailed knowledge of their insecticidal mechanisms. However, despite the wealth of knowledge on the insecticidal effectiveness of nanomaterials against insect pests [164], exact information on their probable mode of action and mechanism in the face of insects is scarce. Mainly, nanomaterials were investigated *in vitro* using bacteria as model microorganisms to ascertain their mode of action [215, 216]. Few studies have been conducted on their possible mechanism of action against insect pests. In the accepted theory, nanomaterials target insects by establishing oxidative stress [217]. Additionally, it is believed that nanomaterials penetrate from the exoskeleton and enter the intracellular region, where they bind to phosphorus from DNA or sulfur from proteins, resulting in the denaturation of nucleic acid and enzymes [218–220]. The insecticidal mechanism of myco-NPs is depicted schematically in Figure 8.9.

8.7.2.2   Role in the Supply of Nutrients and Plant Growth Promotion

Current agriculture systems rely heavily on chemical fertilizers. Although synthetic chemical inputs have lost their efficiency in today's agriculture system, fertilizers are vital for plant growth and provide nutrients to plants, and it is impossible to remove them. The overuse of chemical and conventional fertilizers has accompanying problems of water pollution, reduced soil fertility, accelerated eutrophication, and greenhouse gas emissions [221, 222]. As a potential alternative to chemical fertilizers, nanofertilizers are mineral nutrients to enhance plant growth and seed germination while also alleviating the adverse effects of commercial fertilizers on the environment [164]. Nanofertilizers are divided into macronutrients and micronutrients, mainly manufactured in two ways. In the first method, nanomaterials are used to encapsulate micro- and macronutrients such as nitrogen (N), copper (Cu), magnesium (Mg), iron (Fe), potassium (K), phosphorus (P), manganese (Mn), and zinc (Zn). Various nanomaterials are used to encapsulate fertilizers to enhance crop uptake and reduce fertilizer outflow. Fertilizers enclosed by NPs are negligibly affected by environmental components such as sunlight, rain, wind, evaporation, weathering, and degradation by microorganisms, so nanomaterials efficiently transport fertilizers into plant cells. Indeed, encapsulation extends the time release of fertilizer, improves the soil's physical and chemical qualities, and makes the soil more fertile [221, 223, 224]. For example, in cultivated coffee plants, absorption of N, P, and K enclosed in chitosan-NPs improved by 17.4%, 16.31%, and 67.50%, respectively, compared to bulk fertilizers [115]. In another study, Aji et al. [114] claimed that carbon dots produced from the dragon fruit (*Hylocereus unddatus*) acted as a carrier of N-P-K fertilizers and had a supplemental role in the stimulation of plant growth of *Ipomoea aquatica*. Elemike et al. [224] asserted that nanofertilizers outperform conventional fertilizers because of their better availability and biocompatibility. Furthermore, the large surface-to-volume ratio, excellent stability, and low toxicity of NPs make nanofertilizers more efficient in releasing their nutrients into the environment [225]. Targeted delivery of nanofertilizers increases their performance and makes them a cost-effective alternative to traditional fertilizers, decreasing application rates [165]. In the second method, nanofertilizers can be produced using reducing agents to reduce elements and production of nanofertilizers. Figure 8.10 depicts the advantages of nanofertilizers schematically.

Using myco-NPs to enhance soil macro- and micronutrients can be an attractive suggestion when using nanotechnology to promote plant growth and boost

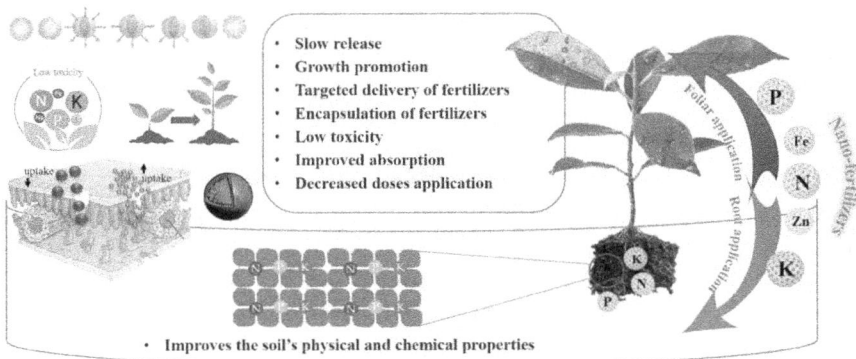

FIGURE 8.10 Schematic depiction of the advantages of nanofertilizers.

(a)   (b)   (c)   (d)

FIGURE 8.11 Effect of zinc nanoparticle (ZnNP) complex on *Vicia faba* seedling growth promotion in Erlenmeyer flask conditions. After 4 days: (a) untreated seeds; (b) treated seeds with ZnNP complex at 20 ppm dose; after 20 days: (c) untreated seeds; (d) treated seeds with ZnNP complex at 20 ppm concentration [196].

seed germination. For example, Moustafa and Taha [196] demonstrated that the mycogenic nano-Zn(II) complex has effective potential in enhancing the growing seedlings of *Vicia faba* in Erlenmeyer flask conditions. Zn(II) complex in the nano-scale biosynthesized using *Trichoderma longibrachiatum* suppressed *Pythium aphanidermatum* damping off. In addition, ZnNPs at 20 ppm increased the weight and length of roots and shoots: shoot and root lengths increased to 180 and 96.5 mm, respectively, compared to controls. ZnNPs also boosted germination rates up to 100% after 72 hours of treatment (Figure 8.11).

Amin et al. [210] biosynthesized SeNPs using metabolites secreted by *Penicillium chrysogenum* as reducing agents. SeNPs can affect plant development either positively or negatively, depending on the dosage utilized. SeNPs can help minimize genotoxicity and electrolyte leakage, enhancing the photosynthetic process and yield productivity at low doses. However, at high concentrations, it can cause oxidative stress, damage plant cells and yield productivity, and disrupt plant cell structure [103, 106, 226]. SeNPs were synthesized using *Penicillium chrysogenum* at concentrations of 5, 10, 15, 20, and 25 ppm. These mycogenic NPs increased that growth development and boosted the content of photosynthetic pigments, carbohydrates, proteins, and proline in plants treated with SeNPs, especially in 20 ppm concentrations in field conditions [210]. Similarly, Nandini et al. [76] asserted that mycogenic SeNPs made from six species of *Trichoderma* spp. can promote the growth of pearl millet. Therefore, nanofertilizers are preferred over conventional inputs and can hold a worthwhile place in agriculture.

8.7.2.3 Myconanoparticle Applications as Remediation

A biological mechanism that recycles waste into another form, called "bioremediation," can be used and reused by other organisms [227–229]. In other words, this mechanism includes eradication, immobilization, and degradation of detoxification in physicochemical waste in the environment, such as oil, heavy metal, pesticide, hydrocarbons, and dyes by microorganisms such as fungi [230].

Mycoremediation (myco-rem) is the application of fungal species to remove hazardous material from the surroundings [231, 232]. Benefits of bioremediation by myco-NPs, are not only economic aspects but also include minimization of biological and chemical toxins, selective to particular metals, no exact nutrient requirements, renewal of biosorbent, and metal recovery [233, 234]. The application of myco-NPs as remediating agents is based on the redox reaction concept in which a neutral electron donor (usually a metal) reduces an electron acceptor. This metal property leads to their potential use as reducing agents for site remediation. Gadd (2009) reported that the diverse metabolism capabilities of fungi play an essential role in the detoxification of organic pollutants and are involved in remediating heavy metals in the environment [235]. Fungi are robust organisms that can tolerate high levels of physicochemical contamination, such as heavy metals [236, 237]. Due to their bioremediation potential, from 1980 to now, fungi have been frequently studied. Among the fungi that are resistant to heavy metals are, for instance, *Fusarium* [238, 239], *Penicillium* [240, 241], *Rhizopus*, and *Aspergillus* [242].

8.8 FUTURE PERSPECTIVES FOR THE APPLICATION OF NANOPARTICLES

For the biological synthesis of NPs, fungi are now known to be a safe, inexpensive, and efficient tool (by both intra- and extracellular methods). Among microorganisms, fungi have been studied due to their compatibility with other microorganisms and also as the handling of fungal biomass and downstream processing is much simpler. Fungal species have successfully synthesized many metallic NPs. Myco-NPs have been successfully used in various fields such as agriculture, health, and the pharmaceutical industry. Myco-rem encompasses the use of fungi in the substantial or partial remediation of particular heavy-metal contaminations in surface and groundwater, industrial wastewater, soil, sediment, and sludge. However, the use of endophytic fungi in NP synthesis has some limitations, such as the necessity for knowledge about the fungus to be employed and its growth parameters, the need for sterile circumstances, and the time required for fungal development and synthesis [35]. The definitive mechanism of myco-NP synthesis is yet to be discovered. So, the application of myco-NPs needs to be further explored before their complete potential can be revealed. Understanding the exact mechanism involved in myco-NPs synthesis will help in developing low-cost techniques for the synthesis and discovery of NPs. Thus, studying different practicalities and reducing agents involved in myco-NP synthesis would help in understanding the fungal system as one of the most efficient biosystems for myco-NP synthesis.

8.9 CONCLUSION AND PROSPECTS

Annually, the frontiers of nanotechnology research and its applications are expanding. With more researchers interested in nanotechnology, much research is now looking into how nanotechnology can be used differently. Today, nanotechnology is well established in the agriculture sector. Nanomaterials are considered smart bullets, targeting specific sites to deliver their active compounds to target sites. Investigation of the interaction of nanomaterials (positive or negative) with plants and microorganisms and the usage of nanomaterials as nano-carriers for controlled release of active substances or the construction of nano-devices is gaining momentum. Many years have been spent developing new methods for synthesizing and characterizing nanomaterials, especially synthesizing using biological methods. In the biosynthesis of nanomaterials, there is an increasing interest in using fungi. Fungi have been proven to be functional bio-manufacturing units that could biosynthesize nanomaterials with a relatively rapid and eco-friendly approach. Non-pathogenic fungi are one of the preferred microorganisms for the biosynthesis of cost-effective nanomaterials due to their ability to produce a higher yield of nanomaterials as a result of metabolite secretion. Thus, nanotechnology based on fungi, called "myconanotechnology," is currently essential science in nanotechnology.

In this chapter, we have attempted to investigate the potential of myco-NPs by discussing their synthesis method, mechanism of synthesis, characteristics, and applications, especially with an emphasis on the role of myconanotechnology in agriculture and pest management. In traditional agriculture, indiscriminate use of conventional agrochemicals has raised concerns for human health and other environmental components. The entrance and release of these agrochemicals into the soil, water, and environment cause primary and secondary effects on the ecosystem. The proper use of agrochemicals in the nanoscale can be a promising solution to this challenge and can be suggested as an alternative strategy. These materials are practical choices against plant pests or act as plant growth promoters. Also, many nano-agrochemicals can act smart and be nano-carriers to deliver the required amount of materials to the target tissue.

Despite the enormous advantages of nanotechnology, some gaps and challenges need to be minimized. The safety of myco-NPs should be confidently revealed. The impact of different factors such as fungi type, growth stage, concentration, exposure time, and culture medium should be accurately assessed. After minimizing these gaps, we need to extend the upcoming horizons and act by filling in the main gaps that are obstacles to our transition from *in vitro* research to field applications.

REFERENCES

1. Wu Q, Miao W-s, Gao H.-j, Hui D. Mechanical properties of nanomaterials: a review. Nanotechnology Reviews, 2020; 9(1): 259–273.
2. Jena B, Ningthoujam R, Pattanayak S, Dash S, Panda MK, Jit BP, Das M, Singh YD, Nanotechnology and its potential application in postharvest technology. *In*: Bio-Nano Interface. 2022; Springer, Singapore. pp. 93–107.
3. Abid N, Khan AM, Shujait S, Chaudhary K, Ikram M, Imran M, Haider J, Khan M, Khan Q, Maqbool M. Synthesis of nanomaterials using various top-down and bottom-up approaches, influencing factors, advantages, and disadvantages: a review. Advances in Colloid and Interface Science, 2021; 102597.
4. Dikmen G, Genç L, Güney G. Advantage and disadvantage in drug delivery systems. Journal of Materials Science and Engineering, 2011; 5(4): 468.
5. Patil AG, Kounaina K, Aishwarya S, Harshitha N, Satapathy P, Hudeda SP, Reddy KR, Alrafas H, Yadav AN, Raghu AV, Zameer F. Myco-nanotechnology for sustainable agriculture: challenges and opportunities. *In*: Recent Trends in Mycological Research: Volume 1: Agricultural and Medical Perspective. A.N. Yadav (Ed.). 2021; Springer International Publishing, Cham. pp. 457–479.

6. Sasidharan S, Raj S, Sonawane S, Pinjari D, Pandit A, Saudagar P. Nanomaterial synthesis: chemical and biological route and applications. *In*: Nanomaterials Synthesis: Design, Fabrication and Applications Micro and Nano Technologies. 2019; Elsevier, USA. pp. 27–51.
7. Shah MA, Pirzada BM, Price G, Shibiru AL, Qurashi A. Applications of nanotechnology in smart textile industry: a critical review. Journal of Advanced Research, 2022; 38: 55–75.
8. Nizami A-S, Rehan M. Towards nanotechnology-based biofuel industry. Biofuel Research Journal, 2018; 5(2): 798–799.
9. Thangadurai D, Sangeetha J, Prasad R. Nanotechnology for Food, Agriculture, and Environment. 2020; Springer, Cham. 405p.
10. Pandey P. Role of nanotechnology in electronics: a review of recent developments and patents. Recent Patents on Nanotechnology, 2022; 16(1): 45–66.
11. Taran M, Safaei M, Karimi N, Almasi A. Benefits and application of nanotechnology in environmental science: an overview. Biointerface Research in Applied Chemistry, 2021; 11(1): 7860–7870.
12. Raina N, Sharma P, Slathia PS, Bhagat D, Pathak AK. Efficiency enhancement of renewable energy systems using nanotechnology. *In*: Nanomaterials and Environmental Biotechnology. 2020; Springer, Cham. pp. 271–297.
13. Leso V, Fontana L, Iavicoli I. Biomedical nanotechnology: occupational views. Nano Today, 2019; 24: 10–14.
14. Prasad M, Lambe UP, Brar B, Shah I, Manimegalai J, Ranjan K, Rao R, Kumar S, Mahant S, Khurana SK. Nanotherapeutics: an insight into healthcare and multidimensional applications in medical sector of the modern world. Biomedicine & Pharmacotherapy, 2018; 97: 1521–1537.
15. Mukhopadhyay SS. Nanotechnology: let the land not be parched. *In*: Nanomaterials Applications for Environmental Matrices. R.F.d. Nascimento, et al. (Eds.). 2019; Elsevier, Amsterdam. pp. 335–353.
16. Gericke M, Pinches A. Biological synthesis of metal nanoparticles. Hydrometallurgy, 2006; 83(1–4): 132–140.
17. Singh T, Jyoti K, Patnaik A, Singh A, Chauhan R, Chandel S. Biosynthesis, characterization and antibacterial activity of silver nanoparticles using an endophytic fungal supernatant of *Raphanus sativus*. Journal of Genetic Engineering and Biotechnology, 2017; 15(1): 31–39.
18. Hassanisadi M, Shahidi Bonjar GH. Plants used in folkloric medicine of Iran are exquisite bio-resources in production of silver nanoparticles. IET Nanobiotechnology, 2017; 11(3): 300–309.
19. Heydari M, Yousefi AR, Nikfarjam N, Rahdar A, Kyzas GZ, Bilal M. Plant-based nanoparticles prepared from protein containing tribenuron-methyl: fabrication, characterization, and application. Chemical and Biological Technologies in Agriculture, 2021; 8(1): 1–11.
20. Salem SS. Bio-fabrication of selenium nanoparticles using baker's yeast extract and its antimicrobial efficacy on food borne pathogens. Applied Biochemistry and Biotechnology, 194(5): 1898–1910.
21. Bahrulolum H, Nooraei S, Javanshir N, Tarrahimofrad H, Mirbagheri VS, Easton AJ, Ahmadian G. Green synthesis of metal nanoparticles using microorganisms and their application in the agrifood sector. Journal of Nanobiotechnology, 2021; 19(1): 1–26.
22. Popli D, Anil V, Subramanyam AB, Namratha MN, Ranjitha VR, Rao SN, Rai RV, Govindappa M. Endophyte fungi, *Cladosporium* species-mediated synthesis of silver nanoparticles possessing in vitro antioxidant, anti-diabetic and

anti-Alzheimer activity. Artificial Cells, Nanomedicine, and Biotechnology, 2018; 46(suppl. 1): 676–683.

23. Parveen K, Banse V, Ledwani L. Green synthesis of nanoparticles: their advantages and disadvantages. *In*: AIP Conference Proceedings. 2016; AIP Publishing.

24. Gahlawat G, Choudhury AR. A review on the biosynthesis of metal and metal salt nanoparticles by microbes. RSC Advances, 2019; 9(23): 12944–12967.

25. Domany El EB, Essam TM, Ahmed AE, Farghali AA. Biosynthesis physico-chemical optimization of gold nanoparticles as anti-cancer and synergetic antimicrobial activity using *Pleurotus ostreatus* fungus. Journal of Applied Pharmaceutical Science, 2018; 8: 119–128.

26. Sreedharan SM, Gupta S, Saxena AK, Singh R. *Macrophomina phaseolina*: microbased biorefinery for gold nanoparticle production. Annals of Microbiology, 2019; 69(4): 435–445.

27. Ma L, Su W, Liu J-X, Zeng X-X, Huang Z, Li W, Liu Z-C, Tang J-X. Optimization for extracellular biosynthesis of silver nanoparticles by *Penicillium aculeatum* Su1 and their antimicrobial activity and cytotoxic effect compared with silver ions. Materials Science and Engineering: C, 2017; 77: 963–971.

28. Rai M, Bridge PD. Applied Mycology. 2009; CABI, Wallingford. 336p.

29. Ningaraju S, Munawer U, Raghavendra VB, Balaji KS, Melappa G, Brindhadevi K, Pugazhendhi A. *Chaetomium globosum* extract mediated gold nanoparticle synthesis and potent anti-inflammatory activity. Analytical Biochemistry, 2021; 612: 113970.

30. Rai M, Gupta I, Bonde S, Ingle P, Shende S, Gaikwad S, Razzaghi-Abyaneh M, Gade A. Industrial applications of nanomaterials produced from *Aspergillus* species. *In*: The Genus *Aspergillus* – Pathogenicity, Mycotoxin Production and Industrial Applications. M Razzaghi-Abyaneh, M Rai (Eds.). 2022; InTechOpen, London. DOI: 10.5772/intechopen.98780

31. Sandhu SS, Shukla H, Shukla S. Biosynthesis of silver nanoparticles by endophytic fungi: its mechanism, characterization techniques and antimicrobial potential. African Journal of Biotechnology,. 2017; 16(14): 683–698.

32. Ealia SAM, Saravanakumar M. A review on the classification, characterisation, synthesis of nanoparticles and their application. *In*: IOP Conference Series: Materials Science and Engineering. 2017; IOP Publishing, UK.

33. Banerjee K, Ravishankar Rai V. A review on mycosynthesis, mechanism, and characterization of silver and gold nanoparticles. BioNanoScience, 2018; 8(1): 17–31.

34. Raghunath A, Perumal E. Metal oxide nanoparticles as antimicrobial agents: a promise for the future. International Journal of Antimicrobial Agents, 2017; 49(2): 137–152.

35. Guilger-Casagrande M, Lima RD. Synthesis of silver nanoparticles mediated by fungi: a review. Frontiers in Bioengineering and Biotechnology, 2019; 7: 287.

36. Chi NTL, Veeraragavan GR, Brindhadevi K, Chinnathambi A, Salmen SH, Alharbi SA, Krishnan R, Pugazhendhi A. Fungi fabrication, characterization, and anticancer activity of silver nanoparticles using metals resistant *Aspergillus niger*. Environmental Research, 2022; 112721.

37. Mistry H, Thakor R, Patil C, Trivedi J, Bariya H. Biogenically proficient synthesis and characterization of silver nanoparticles employing marine procured fungi *Aspergillus brunneoviolaceus* along with their antibacterial and antioxidative potency. Biotechnology Letters, 2021; 43(1): 307–316.

38. Othman AM, Elsayed MA, Al-Balakocy NG, Hassan MM, Elshafei AM. Biosynthesized silver nanoparticles by *Aspergillus terreus* NRRL265 for imparting durable antimicrobial finishing to polyester cotton blended fabrics: statistical optimization,

characterization, and antitumor activity evaluation. Biocatalysis and Agricultural Biotechnology, 2021; 31: 101908.

39. Shahzad A, Saeed H, Iqtedar M, Hussain SZ, Kaleem A, Abdullah R, Sharif S, Naz S, Saleem F, Aihetasham A. Size-controlled production of silver nanoparticles by *Aspergillus fumigatus* BTCB10: likely antibacterial and cytotoxic effects. Journal of Nanomaterials, 2019; 2019: 1–14.

40. Ottoni CA, Simões MF, Fernandes S, Dos Santos JG, Da Silva ES, de Souza RFB, Maiorano AE. Screening of filamentous fungi for antimicrobial silver nanoparticles synthesis. AMB Express, 2017; 7(1): 1–10.

41. Rodríguez-Serrano C, Guzmán-Moreno J, Ángeles-Chávez C, Rodríguez-González V, Ortega-Sigala JJ, Ramírez-Santoyo RM, Vidales-Rodríguez LE. Biosynthesis of silver nanoparticles by *Fusarium scirpi* and its potential as antimicrobial agent against uropathogenic *Escherichia coli* biofilms. Plos One, 2020; 15(3): e0230275.

42. El-Sayed E-SR, Abdelhakim HK, Zakaria Z. Extracellular biosynthesis of cobalt ferrite nanoparticles by *Monascus purpureus* and their antioxidant, anticancer and antimicrobial activities: yield enhancement by gamma irradiation. Materials Science and Engineering: C, 2020; 107: 110318.

43. Ajah A, Hassan AS, Aja HA. Extracellular biosynthesis of silver nanoparticles using *Fusarium graminearum* and their antimicrobial activity. Journal of Global Pharma Technology, 2018; 10: 683–689.

44. Ghabooli A, Mirzaei S. Biosynthesis of silver nanoparticles using *Aspergillus flavus* and investigation of some effective factors on its production. Biological Journal of Microorganism, 2018; 7(27): 81–94.

45. Khalil NM, Abd El- Ghany MN, Rodríguez-Couto S. Antifungal and anti-mycotoxin efficacy of biogenic silver nanoparticles produced by *Fusarium chlamydosporum* and *Penicillium chrysogenum* at non-cytotoxic doses. Chemosphere, 2019; 218: 477–486.

46. Rai M, Bonde S, Golinska P, Trzcińska-Wencel J, Gade A, Abd-Elsalam K, Shende S, Gaikwad S, Ingle A. *Fusarium* as a novel fungus for the synthesis of nanoparticles: mechanism and applications. Journal of Fungi, 2021; 7(2): 139.

47. Yassin MA, Elgorban AM, El-Samawaty AE-RM, Almunqedhi BM. Biosynthesis of silver nanoparticles using *Penicillium verrucosum* and analysis of their antifungal activity. Saudi Journal of Biological Sciences, 2021; 28(4): 2123–2127.

48. Naveen KV, Sathiyaseelan A, Mariadoss AVA, Xiaowen H, Saravanakumar K, Wang M-H. Fabrication of mycogenic silver nanoparticles using endophytic fungal extract and their characterization, antibacterial and cytotoxic activities. Inorganic Chemistry Communications, 2021; 128: 108575.

49. Hamad M. Biosynthesis of silver nanoparticles by fungi and their antibacterial activity. International Journal of Environmental Science and Technology, 2019; 16(2): 1015–1024.

50. Nayak B, Nanda A, Prabhakar V. Biogenic synthesis of silver nanoparticle from wasp nest soil fungus, *Penicillium italicum* and its analysis against multi drug resistance pathogens. Biocatalysis and Agricultural Biotechnology, 2018; 16: 412–418.

51. Ameen F, Al-Homaidan AA, Al-Sabri A, Almansob A, AlNadhari S. Anti-oxidant, anti-fungal and cytotoxic effects of silver nanoparticles synthesized using marine fungus *Cladosporium halotolerans*. Applied Nanoscience, 2021; 1–9.

52. Guilger-Casagrande M, Germano-Costa T, Pasquoto-Stigliani T, Fraceto LF, Lima RD. Biosynthesis of silver nanoparticles employing *Trichoderma harzianum* with enzymatic stimulation for the control of *Sclerotinia sclerotiorum*. Scientific Reports, 2019; 9(1): 1–9.

53. Mirzaei S, Ghabooli A, Mirzaei M. *Botrytis cinerea*, one of the most destructive plant pathogens, as a potent to produce silver nanoparticles. International Journal of Nanoscience and Nanotechnology, 2020; 16(4): 243–248.

54. Saxena J, Sharma PK, Sharma MM, Singh A. Process optimization for green synthesis of silver nanoparticles by *Sclerotinia sclerotiorum* MTCC 8785 and evaluation of its antibacterial properties. SpringerPlus, 2016; 5(1): 1–10.

55. Xue B, He D, Gao S, Wang D, Yokoyama K, Wang L. Biosynthesis of silver nanoparticles by the fungus *Arthroderma fulvum* and its antifungal activity against genera of *Candida*, *Aspergillus* and *Fusarium*. International Journal of Nanomedicine, 2016; 11: 1899.

56. Amerasan D, Nataraj T, Murugan K, Panneerselvam C, Madhiyazhagan P, Nicoletti M, Benelli G. Myco-synthesis of silver nanoparticles using *Metarhizium anisopliae* against the rural malaria vector *Anopheles culicifacies* Giles (Diptera: Culicidae). Journal of Pest Science, 2016; 89(1): 249–256.

57. Santos TS, Silva TM, Cardoso JC, de Albuquerque-Júnior RL, Zielinska A, Souto EB, Severino P, Mendonça MdC. Biosynthesis of silver nanoparticles mediated by entomopathogenic fungi: antimicrobial resistance, nanopesticides, and toxicity. Antibiotics, 2021; 10(7): 852.

58. Bhardwaj K, Sharma A, Tejwan N, Bhardwaj S, Bhardwaj P, Nepovimova E, Shami A, Kalia A, Kumar A, Abd-Elsalam KA, Kuča K. *Pleurotus* macrofungi-assisted nanoparticle synthesis and its potential applications: a review. Journal of Fungi (Basel), 2020; 6(4): 351–372.

59. Krishna G, Srileka V, Charya MS, Serea ESA, Shalan AE. Biogenic synthesis and cytotoxic effects of silver nanoparticles mediated by white rot fungi. Heliyon, 2021; 7(3): e06470.

60. Osorio-Echavarría J, Osorio-Echavarría J, Ossa-Orozco CP, Gómez-Vanegas NA. Synthesis of silver nanoparticles using white-rot fungus anamorphous *Bjerkandera* sp. R1: influence of silver nitrate concentration and fungus growth time. Scientific Reports, 2021; 11(1): 1–14.

61. Ameen F, Al-Maary KS, Almansob A, AlNadhari S. Antioxidant, antibacterial and anticancer efficacy of *Alternaria chlamydospora*-mediated gold nanoparticles. Applied Nanoscience, 2022; 1–8.

62. Clarance P, Luvankar B, Sales J, Khusro A, Agastian P, Tack J-C, Al Khulaifi MM, Al-Shwaiman HA, Elgorban AM, Syed A. Green synthesis and characterization of gold nanoparticles using endophytic fungi *Fusarium solani* and its *in-vitro* anticancer and biomedical applications. Saudi Journal of Biological Sciences, 2020; 27(2): 706–712.

63. Munawer U, Raghavendra VB, Ningaraju S, Krishna KL, Ghosh AR, Melappa G, Pugazhendhi A. Biofabrication of gold nanoparticles mediated by the endophytic *Cladosporium* species: photodegradation, in vitro anticancer activity and in vivo antitumor studies. International Journal of Pharmaceutics, 2020; 588: 119729.

64. Naimi-Shamel N, Pourali P, Dolatabadi S. Green synthesis of gold nanoparticles using *Fusarium oxysporum* and antibacterial activity of its tetracycline conjugant. Journal de Mycologie Medicale, 2019; 29(1): 7–13.

65. Molnár Z, Bódai V, Szakacs G, Erdélyi B, Fogarassy Z, Sáfrán G, Varga T, Kónya Z, Tóth-Szeles E, Szűcs R. Green synthesis of gold nanoparticles by thermophilic filamentous fungi. Scientific Reports, 2018; 8(1): 1–12.

66. Bhargava A, Jain N, Khan MA, Pareek V, Dilip RV, Panwar J. Utilizing metal tolerance potential of soil fungus for efficient synthesis of gold nanoparticles with superior catalytic activity for degradation of rhodamine B. Journal of Environmental Management, 2016; 183: 22–32.

67. Soltani Nejad M, Najafabadi NS, Aghighi S, Pakina E, Zargar M. Evaluation of *Phoma* sp. biomass as an endophytic fungus for synthesis of extracellular gold nanoparticles with antibacterial and antifungal properties. Molecules, 2022; 27(4): 1181.

68. Sreedharan SM, Gupta S, Saxena AK, Singh R. *Macrophomina phaseolina*: microbased biorefinery for gold nanoparticle production. Annals of Microbiology, 2019; 69: 435–445.

69. Al-Khattaf FS. Gold and silver nanoparticles: green synthesis, microbes, mechanism, factors, plant disease management and environmental risks. Saudi Journal of Biological Sciences, 2021; 28(6): 3624–3631.

70. Subashini G, Bhuvaneswari S. Nanoparticles from fungi (myconanoparticles). *In*: Fungi and their Role in Sustainable Development: Current Perspectives. P. Gehlot, J. Singh (Eds.). 2018; Springer Singapore, Singapore. pp. 753–779.

71. Liang X, Perez MAM-J, Nwoko KC, Egbers P, Feldmann J, Csetenyi L, Gadd GM. Fungal formation of selenium and tellurium nanoparticles. Applied Microbiology and Biotechnology, 2019; 103(17): 7241–7259.

72. Espinosa-Ortiz EJ, Rene ER, Guyot F, van Hullebusch ED, Lens PN. Biomineralization of tellurium and selenium-tellurium nanoparticles by the white-rot fungus *Phanerochaete chrysosporium*. International Biodeterioration & Biodegradation, 2017; 124: 258–266.

73. Pieła A, Żymańczyk-Duda E, Brzezińska-Rodak M, Duda M, Grzesiak J, Saeid A, Mironiuk M, Klimek-Ochab M. Biogenic synthesis of silica nanoparticles from corn cobs husks. Dependence of the productivity on the method of raw material processing. Bioorganic Chemistry, 2020; 99: 103773.

74. Vetchinkina E, Loshchinina E, Kurskyi V, Nikitina V. Biological synthesis of selenium and germanium nanoparticles by xylotrophic basidiomycetes. Applied Biochemistry and Microbiology, 2016; 52(1): 87–97.

75. Zielonka A, Żymańczyk-Duda E, Brzezińska-Rodak M, Duda M, Grzesiak J, Klimek-Ochab M. Nanosilica synthesis mediated by *Aspergillus parasiticus* strain. Fungal Biology, 2018; 122(5): 333–344.

76. Nandini B, Hariprasad P, Prakash HS, Shetty HS, Geetha N. Trichogenic-selenium nanoparticles enhance disease suppressive ability of *Trichoderma* against downy mildew disease caused by *Sclerospora graminicola* in pearl millet. Scientific Reports, 2017; 7(1): 1–11.

77. Joshi SM, De Britto S, Jogaiah S, Ito S-i. Mycogenic selenium nanoparticles as potential new generation broad spectrum antifungal molecules. Biomolecules, 2019; 9(9): 419.

78. García RÁ-F, Corte-Rodríguez M, Macke M, LeBlanc K, Mester Z, Montes-Bayón M, Bettmer J. Addressing the presence of biogenic selenium nanoparticles in yeast cells: analytical strategies based on ICP-TQ-MS. Analyst, 2020; 145(4): 1457–1465.

79. Bafghi MH, Darroudi M, Zargar M, Zarrinfar H, Nazari R. Biosynthesis of selenium nanoparticles by *Aspergillus flavus* and *Candida albicans* for antifungal applications. Micro & Nano Letters, 2021; 16(14): 656–669.

80. El-Sayed E-SR, Abdelhakim HK, Ahmed AS. Solid-state fermentation for enhanced production of selenium nanoparticles by gamma-irradiated *Monascus purpureus* and their biological evaluation and photocatalytic activities. Bioprocess and Biosystems Engineering, 2020; 43(5): 797–809.

81. Sun Y, Shi Y, Jia H, Ding H, Yue T, Yuan Y. Biosynthesis of selenium nanoparticles of *Monascus purpureus* and their inhibition to *Alicyclobacillus acidoterrestris*. Food Control, 2021; 130: 108366.

82. Amin BH, Ahmed HY, El Gazzar EM, Badawy MM. Enhancement the mycosynthesis of selenium nanoparticles by using gamma radiation. Dose-Response, 2021; 19(4): 15593258211059323.

83. Vetchinkina E, Loshchinina E, Kupryashina M, Burov A, Nikitina V. Shape and size diversity of gold, silver, selenium, and silica nanoparticles prepared by green synthesis using fungi and bacteria. Industrial & Engineering Chemistry Research, 2019; 58(37): 17207–17218.

84. Hu D, Yu S, Yu D, Liu N, Tang Y, Fan Y, Wang C, Wu A. Biogenic *Trichoderma harzianum*-derived selenium nanoparticles with control functionalities originating from diverse recognition metabolites against phytopathogens and mycotoxins. Food Control, 2019; 106: 106748.

85. Zhang H, Zhou H, Bai J, Li Y, Yang J, Ma Q, Qu Y. Biosynthesis of selenium nanoparticles mediated by fungus *Mariannaea* sp. HJ and their characterization. Colloids and Surfaces A: Physicochemical and Engineering Aspects, 2019; 571: 9–16.

86. Elsoud MMA, Al-Hagar OE, Abdelkhalek ES, Sidkey N. Synthesis and investigations on tellurium myconanoparticles. Biotechnology Reports, 2018; 18: e00247.

87. Rose GK, Soni R, Rishi P, Soni SK. Optimization of the biological synthesis of silver nanoparticles using *Penicillium oxalicum* GRS-1 and their antimicrobial effects against common food-borne pathogens. Green Processing and Synthesis, 2019; 8(1): 144–156.

88. Phanjom P, Ahmed G. Effect of different physicochemical conditions on the synthesis of silver nanoparticles using fungal cell filtrate of *Aspergillus oryzae* (MTCC no. 1846) and their antibacterial effect. Advances in Natural Sciences: Nanoscience and Nanotechnology, 2017; 8(4): 045016.

89. Azmath P, Baker S, Rakshith D, Satish S. Mycosynthesis of silver nanoparticles bearing antibacterial activity. Saudi Pharmaceutical Journal, 2016; 24(2): 140–146.

90. Ameen F. Optimization of the synthesis of fungus-mediated bi-metallic Ag-Cu nanoparticles. Applied Sciences, 2022; 12(3): 1384.

91. Khandel P, Shahi SK. Microbes mediated synthesis of metal nanoparticles: current status and future prospects. International Journal of Nanomaterial Biostructure, 2016; 6(1): 1–24.

92. Rónavári A, Igaz N, Adamecz DI, Szerencsés B, Molnar C, Kónya Z, Pfeiffer I, Kiricsi M. Green silver and gold nanoparticles: biological synthesis approaches and potentials for biomedical applications. Molecules, 2021; 26(4): 844.

93. Mukherji S, Bharti S, Shukla G, Mukherji S. Synthesis and characterization of size- and shape-controlled silver nanoparticles. *In*: Volume 1 B Metallic Nanomaterials (Part B). 2018; De Gruyter, Berlin. pp. 1–116.

94. Hietzschold S, Walter A, Davis C, Taylor AA, Sepunaru L. Does nitrate reductase play a role in silver nanoparticle synthesis? Evidence for NADPH as the sole reducing agent. ACS Sustainable Chemistry & Engineering, 2019; 7(9): 8070–8076.

95. Lan Chi NT, Veeraragavan GR, Brindhadevi K, Chinnathambi A, Salmen SH, Alharbi SA, Krishnan R, Pugazhendhi A. Fungi fabrication, characterization, and anticancer activity of silver nanoparticles using metals resistant *Aspergillus niger*. Environmenal Research, 2022; 208: 112721.

96. Mistry H, Thakor R, Patil C, Trivedi J, Bariya H. Biogenically proficient synthesis and characterization of silver nanoparticles employing marine procured fungi *Aspergillus brunneoviolaceus* along with their antibacterial and antioxidative potency. Biotechnology Letters, 2021; 43(1): 307–316.

97. Othman AM, Elsayed MA, Al-Balakocy NG, Hassan MM, Elshafei AM. Biosynthesized silver nanoparticles by *Aspergillus terreus* NRRL265 for imparting durable antimicrobial finishing to polyester cotton blended fabrics: statistical optimization,

characterization, and antitumor activity evaluation. Biocatalysis and Agricultural Biotechnology, 2021; 31:101908.

98. Ghabooli A, Mirzaei S. Biosynthesis of silver nanoparticles using *Aspergillus flavus* and investigation of some effective factors in its production. Biological Journal of Microorganism, 2018; 7 (27): 81–94.

99. Farrag HMM, Mostafa FAAM, Mohamed ME, Huseein EAM. Green biosynthesis of silver nanoparticles by *Aspergillus niger* and its antiamoebic effect against *Allovahlkampfia spelaea* trophozoite and cyst. Experimental Parasitology, 2020; 219: 108031. doi: 10.1016/j.exppara.2020.1080312020.

100. Velhal SG, Kulkarni SD, Latpate RV. Fungal mediated silver nanoparticle synthesis using robust experimental design and its application in cotton fabric. International Nano Letters, 2016; 6: 257–264.

101. Rodríguez-Serrano C, Guzmán-Moreno J, Ángeles-Chávez C, Rodríguez-González V, Ortega-Sigala JJ, Ramírez-Santoyo RM, Vidales-Rodríguez LE. Biosynthesis of silver nanoparticles by *Fusarium scirpi* and its potential as antimicrobial agent against uropathogenic *Escherichia coli* biofilms. PLoS One, 2020; 15(3): 1–20.

102. Chatterjee S, Mahanty S, Das P, Chaudhuri P, Das S. Biofabrication of iron oxide nanoparticles using manglicolous fungus *Aspergillus niger* BSC-1 and removal of Cr (VI) from aqueous solution. Chemical Engineering Journal, 2020; 385: 123790.

103. Molnár Á, Kolbert Z, Kéri K, Feigl G, Ördög A, Szőllősi R, Erdei L. Selenite-induced nitro-oxidative stress processes in *Arabidopsis thaliana* and *Brassica juncea*. Ecotoxicology and Environmental Safety, 2018; 148: 664–674.

104. Khalil NM, Abd El-Ghany MN, Rodríguez-Couto S. Antifungal and anti-mycotoxin efficacy of biogenic silver nanoparticles produced by *Fusarium chlamydosporum* and *Penicillium chrysogenum* at non-cytotoxic doses. Chemosphere, 2019; 218: 477–486.

105. Yassin MA, Elgorban AM, El-Samawaty AERMA, Almunqedhi BMA. Biosynthesis of silver nanoparticles using *Penicillium verrucosum* and analysis of their antifungal activity. Saudi Journal of Biological Sciences, 2021; 28(4): 2123–2127.

106. Zohra E, Ikram M, Omar AA, Hussain M, Satti SH, Raja NI, Ehsan M. Potential applications of biogenic selenium nanoparticles in alleviating biotic and abiotic stresses in plants: a comprehensive insight on the mechanistic approach and future perspectives. Green Processing and Synthesis, 2021; 10(1): 456–475.

107. Ma L, Su W, Liu J-X, Zeng X-X, Huang Z, Li W, Liu Z-C, Tang J-X. Optimization for extracellular biosynthesis of silver nanoparticles by *Penicillium aculeatum* Su1 and their antimicrobial activity and cytotoxic effect compared with silver ions. Materials Science & Engineering. C, Materials for Biological Applications, 2017; 77: 963–971.

108. Hamad MT. Biosynthesis of silver nanoparticles by fungi and their antibacterial activity. International Journal of Environmental Science and Technology, 2019; 16: 1015–1024.

109. Nayak BK, Nanda A, Prabhakar V. Biogenic synthesis of silver nanoparticle from wasp nest soil fungus, *Penicillium italicum* and its analysis against multi drug resistance pathogens. Biocatalysis and Agricultural Biotechnology, 2018; 16: 412–418.

110. Azmath P, Baker S, Rakshith D, Satish S. Mycosynthesis of silver nanoparticles bearing antibacterial activity. Saudi Pharmaceutical Journal, 2016; 24(2): 140–146.

111. Singh T, Jyoti K, Patnaik A, Singh A, Chauhan R, Chandel SS. Biosynthesis, characterization and antibacterial activity of silver nanoparticles using an endophytic fungal supernatant of *Raphanus sativus*. Journal of Genetic Engineering and Biotechnology, 2017; 15: 31–39.

112. Xue B, He D, Gao S, Wang D, Yokoyama K, Wang LLB-X. Biosynthesis of silver nanoparticles by the fungus *Arthroderma fulvum* and its antifungal activity against

genera of *Candida, Aspergillus* and *Fusarium*. International Journal of Nanomedicine, 2016; 11: 1899–1906.

113. Yan S, Ren BY, Shen J. Nanoparticle-mediated double-stranded RNA delivery system: a promising approach for sustainable pest management. Insect Science, 2021; 28(1): 21–34.

114. Aji MP, Sholikhah L, Silmi FI, Permatasari HA, Rahmawati I, Priyanto A, Nuryadin BW. Carbon dots from dragonfruit peels as growth-enhancer on *Ipomoea aquatica* vegetable cultivation. Advances in Natural Sciences: Nanoscience and Nanotechnology, 2020; 11(3): 035005.

115. Ha NMC, Nguyen TH, Wang S-L, Nguyen AD. Preparation of NPK nanofertilizer based on chitosan nanoparticles and its effect on biophysical characteristics and growth of coffee in green house. Research on Chemical Intermediates, 2019; 45(1): 51–63.

116. Ahmad SA, Das SS, Khatoon A, Ansari MT, Afzal M, Hasnain MS, Nayak AK. Bactericidal activity of silver nanoparticles: a mechanistic review. Materials Science for Energy Technologies, 2020; 3: 756–769.

117. Bahrulolum H, Nooraei S, Javanshir N, Tarrahimofrad H, Mirbagheri VS, Easton AJ, Ahmadian G. Green synthesis of metal nanoparticles using microorganisms and their application in the agrifood sector. Journal of Nanobiotechnology, 2021; 19: 1–26.

118. Domany El EB, Essam TM, Ahmed AE, Farghali AA. Biosynthesis physico-chemical optimization of gold nanoparticles as anti-cancer and synergetic antimicrobial activity using *Pleurotus ostreatus* fungus. Journal of Applied Pharmaceutical Science, 2018; 8(5): 119–128.

119. Ningaraju S, Munawer U, Raghavendra VB, Balaji KS, Melappa G, Brindhadevi K, Pugazhendhi A. *Chaetomium globosum* extract mediated gold nanoparticle synthesis and potent anti-inflammatory activity. Analytical Biochemistry, 2021; 612: 113970.

120. Abbas A, Naz SS, Syed SA. Antimicrobial activity of silver nanoparticles (AgNPs) against *Erwinia carotovora* subsp. *atroseptica* and *Alternaria alternata*. Pakistani Journal of Agricultural Sciences, 2019; 56(1): 113–117.

121. El Sayed MT, El-Sayed ASA. Biocidal activity of metal nanoparticles synthesized by *Fusarium solani* against multidrug-resistant bacteria and mycotoxigenic fungi. Journal of Microbiology and Biotechnology, 2020; 30: 226–236.

122. Noor S, Shah Z, Javed A, Ali A, Hussain B, Zafar S, Ali H, Muhammad SA. A fungal based synthesis method for copper nanoparticles with the determination of anticancer, antidiabetic and antibacterial activities. Journal of Microbiological Methods, 2020; 174: 105966.

123. Li Q, Gadd GM. Biosynthesis of copper carbonate nanoparticles by ureolytic fungi. Applied Microbiology and Biotechnology, 2017; 101: 7397–7407.

124. Gupta K, Chundawat TS. Zinc oxide nanoparticles synthesized using *Fusarium oxysporum* to enhance bioethanol production from rice-straw. Biomass and Bioenergy, 2020; 143: 105840.

125. Sharma JL, Dhayal V, Sharma RK. White-rot fungus mediated green synthesis of zinc oxide nanoparticles and their impregnation on cellulose to develop environmental friendly antimicrobial fibers. 3 Biotech, 2021; 11(6): 1–10.

126. Kumar RV, Vinoth S, Baskar V, Arun M, Gurusaravanan P. Synthesis of zinc oxide nanoparticles mediated by *Dictyota dichotoma* endophytic fungi and its photocatalytic degradation of fast green dye and antibacterial applications. South African Journal of Botany, 2022; 1–8.

127. Mani VM, Nivetha S, Sabarathinam S, Barath S, Das MPA, Basha S, Elfasakhany A, Pugazhendhi A. Multifunctionalities of mycosynthesized zinc oxide nanoparticles (ZnONPs) from *Cladosporium tenuissimum* FCBGr: antimicrobial additives for

paints coating, functionalized fabrics and biomedical properties. Progress in Organic Coatings, 2022; 163: 106650.

128. Ammar HA, Alghazaly MS, Assem Y, Abou Zeid AA. Bioengineering and optimization of PEGylated zinc nanoparticles by simple green method using *Monascus purpureus*, and their powerful antifungal activity against the most famous plant pathogenic fungi, causing food spoilage. Environmental Nanotechnology, Monitoring & Management, 2021; 16: 100543.

129. Kumaravel J, Lalitha K, Arunthirumeni M, Shivakumar MS. Mycosynthesis of bimetallic zinc oxide and titanium dioxide nanoparticles for control of *Spodoptera frugiperda*. Pesticide Biochemistry and Physiology, 2021; 178: 104910.

130. Arya S, Sonawane H, Math S, Tambade P, Chaskar M, Shinde D. Biogenic titanium nanoparticles (TiO$_2$NPs) from *Trichoderma citrino viride* extract: synthesis, characterization and antibacterial activity against extremely drug-resistant *Pseudomonas aeruginosa*. International Nano Letters, 2021; 11: 35–42.

131. Peiris MMK, Guansekera TDCP, Jayaweera PM, Fernando SSN. TiO$_2$ nanoparticles from baker's yeast: a potent antimicrobial. Journal of Microbiology and Biotechnology, 2018; 28(10): 1664–1670.

132. Suryavanshi P, Pandit R, Gade A, Derita M, Zachino S, Rai M. *Colletotrichum* sp. mediated synthesis of sulphur and aluminium oxide nanoparticles and its in vitro activity against selected food-borne pathogens. LWT Food Science and Technology, 2017; 81: 188–194.

133. Alsaggaf MS, Elbaz AF, Badawy SE, Moussa SH. Anticancer and antibacterial activity of cadmium sulfide nanoparticles by *Aspergillus niger*. Advances in Polymer Technology, 2020; 1–13.

134. Calvo-Olvera A, De Donato-Capote M, Pool H, Rojas-Avelizapa NG. In vitro toxicity assessment of fungal-synthesized cadmium sulfide quantum dots using bacteria and seed germination models. Journal of Environmental Science and Health, Part A: Toxic/Hazardous Substances and Environmental Engineering, 2021; 56(6): 713–722.

135. Vijayanandan AS, Balakrishnan RM. Biosynthesis of cobalt oxide nanoparticles using endophytic fungus *Aspergillus nidulans*. Journal of Environmental Management, 2018; 218: 442–450.

136. Golnaraghi Ghomi AR, Mohammadi-Khanaposhti M, Vahidi H, Kobarfard F, Reza MAS, Barabadi H. Fungus-mediated extracellular biosynthesis and characterization of zirconium nanoparticles using standard *Penicillium* species and their preliminary bactericidal potential: a novel biological approach to nanoparticle synthesis. Iranian Journal of Pharmaceutical Research, 2019; 18: 2101–2110.

137. Gholami-Shabani M, Sotoodehnejadnematalahi F, Shams-Ghahfarokhi M, Eslamifar A, Razzaghi-Abyaneh M. Platinum nanoparticles as potent anticancer and antimicrobial agent: green synthesis, physical characterization, and in-vitro biological activity. Journal of Cluster Science, 2022; https://doi.org/10.1007/s10876-022-02225-6.

138. da Silva BL, Caetano BL, Chiari-Andréo BG, Pietro RCLR, Chiavacci LA. Increased antibacterial activity of ZnO nanoparticles: influence of size and surface modification. Colloids and Surfaces B: Biointerfaces, 2019; 177: 440–447.

139. Bafghi MH, Darroudi, M, Zargar M, Zarrinfar H, Nazari R. Biosynthesis of selenium nanoparticles by *Aspergillus flavus* and *Candida albicans* for antifungal applications. Micro & Nano Letters, 2021; 16: 656–669.

140. Álvarez-Fernández García R, Corte-Rodríguez M, Macke M, Leblanc KL, Mester Z, Montes-Bayón M, Bettmer J. Addressing the presence of biogenic selenium nanoparticles in yeast cells: analytical strategies based on ICP-TQ-MS. Analyst, 2020; 145: 1457–1465.

141. Abo Elsoud MM, Al-Hagar OEA, Abdelkhalek ES, Sidkey NM. Synthesis and investigations on tellurium myconanoparticles. Biotechnology Reports, 2018; 18: e00247.

142. Akhtar I, Iqbal Z, Saddiqe Z. Nanotechnology in pest management. *In*: Nanoagronomy. 2020; Springer, Cham. pp. 69–83.

143. Zielonka A, Żymańczyk-Duda E, Brzezińska-Rodak M, Duda M, Grzesiak J, Klimek-Ochab M. Nanosilica synthesis mediated by *Aspergillus parasiticus* strain. Fungal Biology, 2018; 122(5): 333–344.

144. Gudikandula K, Vadapally P, Charya MS. Biogenic synthesis of silver nanoparticles from white rot fungi: their characterization and antibacterial studies. OpenNano, 2017; 2: 64–78.

145. Fatima I, Rahdar A, Sargazi S, Barani M, Hassanisaadi M, Thakur VK. Quantum dots: synthesis, antibody conjugation, and HER2-receptor targeting for breast cancer therapy. Journal of Functional Biomaterials, 2021; 12(4): 75.

146. Mohammadzadeh V, Barani M, Amiri MS, Yazdi MET, Hassanisaadi M, Rahdar A, Varma RS. Applications of plant-based nanoparticles in nanomedicine: a review. Sustainable Chemistry and Pharmacy, 2022; 25: 100606.

147. Sargazi S, Hosseinikhah SM, Zargari F, Chauhana NPS, Hassanisaadi M, Amani S. pH-responsive cisplatin-loaded niosomes: synthesis, characterization, cytotoxicity study and interaction analyses by simulation methodology. Nanofabrication, 2020; 6(1): 1–15.

148. Sargazi S, Laraib U, Er S, Rahdar A, Hassanisaadi M, Zafar MN, Díez-Pascual AM, Bilal M. Application of green gold nanoparticles in cancer therapy and diagnosis. Nanomaterials, 2022; 12(7): 1102.

149. Morones J, Elechiguerra J, Camacho A, Holt K, Kouri JB, Ramirez JT, Yacaman MJ. The bactericidal effect of silver nanoparticles. Nanotechnology, 2005; 16(10): 2346–2353.

150. Rai M, Bridge PD. Applied Mycology. Wallingford; CABI. 2009; 318p. doi.10.1079/9781845935344.0258

151. Baker S, Satish S. Endophytes: toward a vision in synthesis of nanoparticle for future therapeutic agents. International Journal of Bio-Inorganic Hybrid Nanomaterials, 2012; 1(2): 67–77.

152. Ashajyothi C, Prabhurajeshwar C, Handral HK. Investigation of antifungal and anti-mycelium activities using biogenic nanoparticles: an eco-friendly approach. Environmental Nanotechnology, Monitoring & Management, 2016; 5: 81–87.

153. Reidy B, Haase A, Luch A, Dawson KA, Lynch I. Mechanisms of silver nanoparticle release, transformation and toxicity: a critical review of current knowledge and recommendations for future studies and applications. Materials, 2013; 6(6): 2295–2350.

154. Fatima F, Verma SR, Pathak N, Bajpai P. Extracellular mycosynthesis of silver nanoparticles and their microbicidal activity. Journal of Global Antimicrobial Resistance, 2016; 7: 88–92.

155. Ahluwalia V, Kumar J, Sisodia R, Shakil NA, Walia S. Green synthesis of silver nanoparticles by *Trichoderma harzianum* and their bio-efficacy evaluation against *Staphylococcus aureus* and *Klebsiella pneumonia*. Industrial Crops and Products, 2014; 55: 202–206.

156. Balakumaran M, Ramachandran R, Kalaichelvan P. Exploitation of endophytic fungus, *Guignardia mangiferae* for extracellular synthesis of silver nanoparticles and their in vitro biological activities. Microbiological Research, 2015; 178: 9–17.

157. Shrivastava S, Dash D. Applying nanotechnology to human health: revolution in bio-medical sciences. Journal of Nanotechnology, 2009; doi.org/10.1155/2009/184702

158. Gade A, Bonde P, Ingle A, Marcato P, Duran N, Rai M. Exploitation of *Aspergillus niger* for synthesis of silver nanoparticles. Journal of Biobased Materials and Bioenergy, 2008; 2(3): 243–247.

159. Bouwmeester H, Dekkers S, Noordam MY, Hagens WI, Bulder AS, Heer C De, Voorde SE Ten, Wijnhoven SW, Marvin HJ, Sips AJ. Review of health safety aspects of nanotechnologies in food production. Regulatory Toxicology and Pharmacology, 2009; 53(1): 52–62.

160. Owolade O, Ogunleti D. Effects of titanium dioxide on the diseases, development and yield of edible cowpea. Journal of Plant Protection Research, 2008; 48(3): 329–335.

161. Nayantara KP. Biosynthesis of nanoparticles using eco-friendly factories and their role in plant pathogenicity: a review. Biotechnology Research and Innovation, 2018; 2: 63–73.

162. Gautam AK, Avasthi S. Myconanoparticles as potential pest control agents. *In*: Nanotechnology for Food, Agriculture, and Environment. 2020; Springer, Cham. pp. 189–226.

163. Feregrino-Perez AA, Magaña-López E, Guzmán C, Esquivel K. A general overview of the benefits and possible negative effects of the nanotechnology in horticulture. Scientia Horticulturae, 2018; 238: 126–137.

164. Hassanisaadi M, Barani M, Rahdar A, Heidary M, Thysiadou A, Kyzas GZ. Role of agrochemical-based nanomaterials in plants: biotic and abiotic stress with germination improvement of seeds. Plant Growth Regulation, 2022; 1–44.

165. Okey-Onyesolu CF, Hassanisaadi M, Bilal M, Barani M, Rahdar A, Iqbal J, Kyzas GZ. Nanomaterials as nanofertilizers and nanopesticides: an overview. Chemistry Select, 2021; 6(33): 8645–8663.

166. Salem SS, Fouda A. Green synthesis of metallic nanoparticles and their prospective biotechnological applications: an overview. Biological Trace Element Research, 2021; 199(1): 344–370.

167. Qu Y, Pei X, Shen W, Zhang X, Wang J, Zhang Z, Li S, You S, Ma F, Zhou J. Biosynthesis of gold nanoparticles by *Aspergillus* sp. WL-Au for degradation of aromatic pollutants. Physica E: Low-dimensional Systems and Nanostructures, 2017; 88: 133–141.

168. Wösten HA. Filamentous fungi for the production of enzymes, chemicals and materials. Current Opinion in Biotechnology, 2019; 59: 65–70.

169. Joshi SA, Salvi SP, Berde CP, Berde VB. Applications of myconanoparticles in remediation: current status and future challenges. *In*: Recent Trends in Mycological Research. 2021; Springer, Cham. pp. 225–239.

170. Prasad R, Kumar V, Prasad KS. Nanotechnology in sustainable agriculture: present concerns and future aspects. African Journal of Biotechnology, 2014; 13(6): 705–713.

171. Gull A, Lone AA, Wani NUI. Biotic and abiotic stresses in plants. *In*: Abiotic and Biotic Stresses in Plants. A.B. de Oliveira (Ed.). 2019; InTechOpen, London. pp. 1–19.

172. Hassanisaadi M, Shahidi Bonjar GH, Hosseinipour A, Abdolshahi R, Ait Barka E, Saadoun I. Biological control of *Pythium aphanidermatum*, the causal agent of tomato root rot by two *Streptomyces* root symbionts. Agronomy, 2021; 11(5): 846.

173. Verma S, Nizam S, Verma PK. Biotic and abiotic stress signaling in plants. *In*: Stress Signaling in Plants: Genomics and Proteomics Perspective. Vol. 1. 2013; Springer, New York. pp. 25–49.

174. Alam T, Khan RAA, Ali A, Sher H, Ullah Z, Ali M. Biogenic synthesis of iron oxide nanoparticles via *Skimmia laureola* and their antibacterial efficacy against bacterial wilt pathogen *Ralstonia solanacearum*. Materials Science and Engineering: C, 2019; 98: 101–108.

175. Saranya S, Selvi A, Babujanarthanam R, Rajasekar A, Madhavan J. Insecticidal activity of nanoparticles and mechanism of action. *In*: Model Organisms to Study Biological Activities and Toxicity of Nanoparticles. 2020; Springer, Singapore. pp. 243–266.

176. Abd-Elsalam KA, Prasad R. Nanobiotechnology Applications in Plant Protection. 2018; Springer, Cham. 394p.

177. Fabiyi OA, Alabi RO, Ansari RA. Nanoparticles' synthesis and their application in the management of phytonematodes: an overview. *In*: Management of Phytonematodes: Recent Advances and Future Challenges. 2020; Springer, Singapore. pp. 125–140.

178. Akther T, Hemalatha S. Mycosilver nanoparticles: synthesis, characterization and its efficacy against plant pathogenic fungi. BioNanoScience, 2019; 9(2): 296–301.

179. Soltani Nejad M, Samandari Najafabadi N, Aghighi S, Pakina E, Zargar M. Evaluation of *Phoma* sp. biomass as an endophytic fungus for synthesis of extracellular gold nanoparticles with antibacterial and antifungal properties. Molecules, 2022; 27(4): 1181.

180. Qin Z, Yue Q, Liang Y, Zhang J, Zhou L, Hidalgo OB, Liu X. Extracellular biosynthesis of biocompatible cadmium sulfide quantum dots using *Trametes versicolor*. Journal of Biotechnology, 2018; 284: 52–56.

181. Chinnaperumal K, Govindasamy B, Paramasivam D, Dilipkumar A, Dhayalan A, Vadivel A, Sengodan K, Pachiappan P. Bio-pesticidal effects of *Trichoderma viride* formulated titanium dioxide nanoparticle and their physiological and biochemical changes on *Helicoverpa armigera* (Hub.). Pesticide Biochemistry and Physiology, 2018; 149: 26–36.

182. Jan H, Gul R, Andleeb A, Ullah S, Shah M, Khanum M, Ullah I, Hano C, Abbasi BH. A detailed review on biosynthesis of platinum nanoparticles (PtNPs), their potential antimicrobial and biomedical applications. Journal of Saudi Chemical Society, 2021; 25(8): 101297.

183. Straub C, Colombi E, McCann HC. Population genomics of bacterial plant pathogens. Phytopathology, 2021; 111(1): 23–31.

184. Mohamed AA, Abu-Elghait M, Ahmed NE, Salem SS. Eco-friendly mycogenic synthesis of ZnO and CuO nanoparticles for in vitro antibacterial, antibiofilm, and antifungal applications. Biological Trace Element Research, 2021; 199(7): 2788–2799.

185. Abdel-Azeem A, Nada AA, O'Donovan A, Thakur VK, Elkelish A. Mycogenic silver nanoparticles from endophytic *Trichoderma atroviride* with antimicrobial activity. Journal of Renewable Materials, 2020; 8(2): 171.

186. Gopinath K, Karthika V, Sundaravadivelan C, Gowri S, Arumugam A. Mycogenesis of cerium oxide nanoparticles using *Aspergillus niger* culture filtrate and their applications for antibacterial and larvicidal activities. Journal of Nanostructure in Chemistry, 2015; 5(3): 295–303.

187. Hassanisaadi M, Bonjar GHS, Rahdar A, Pandey S, Hosseinipour A, Abdolshahi R. Environmentally safe biosynthesis of gold nanoparticles using plant water extracts. Nanomaterials, 2021; 11(8): 2033.

188. Gautam N, Salaria N, Thakur K, Kukreja S, Yadav N, Yadav R, Goutam U. Green silver nanoparticles for phytopathogen control. Proceedings of the National Academy of Sciences, India Section B: Biological Sciences, 2020; 90(2): 439–446.

189. Zaki SA, Ouf SA, Abd-Elsalam KA, Asran AA, Hassan MM, Kalia A, Albarakaty FM. Trichogenic silver-based nanoparticles for suppression of fungi involved in damping-off of cotton seedlings. Microorganisms, 2022; 10(2): 344.

190. Khoei NS, Lampis S, Zonaro E, Yrjälä K, Bernardi P, Vallini G. Insights into selenite reduction and biogenesis of elemental selenium nanoparticles by two environmental isolates of *Burkholderia fungorum*. New Biotechnology, 2017; 34: 1–11.
191. Khiralla GM, El-Deeb BA. Antimicrobial and antibiofilm effects of selenium nanoparticles on some foodborne pathogens. LWT-Food Science and Technology. 2015; 63(2): 1001–1007.
192. Vinković Vrček I. Selenium nanoparticles: biomedical applications. *In*: Selenium. B. Michalke (Ed.). 2018; Springer International, Cham. pp. 393–412.
193. Vrandečić K, Ćosić J, Ilić J, Ravnjak B, Selmani A, Galić E, Pem B, Barbir R, Vinković Vrček I, Vinković T. Antifungal activities of silver and selenium nanoparticles stabilized with different surface coating agents. Pest Management Science, 2020; 76(6): 2021–2029.
194. Ferro C, Florindo HF, Santos HA. Selenium nanoparticles for biomedical applications: from development and characterization to therapeutics. Advanced Healthcare Materials, 2021; 10(16): 2100598.
195. Joshi SM, De Britto S, Jogaiah S. Myco-engineered selenium nanoparticles elicit resistance against tomato late blight disease by regulating differential expression of cellular, biochemical and defense responsive genes. Journal of Biotechnology, 2021; 325: 196–206.
196. Moustafa S, Taha RH. Mycogenic nano-complex for plant growth promotion and bio-control of *Pythium aphanidermatum*. Plants, 2021; 10(9): 1858.
197. Shankar Naik B. Nanoparticles from endophytic fungi and their efficacy in biological control. Nanotechnology for Food, Agriculture, and Environment, 2020; 161–179.
198. Khorrami S, Zarrabi A, Khaleghi M, Danaei M, Mozafari M. Selective cytotoxicity of green synthesized silver nanoparticles against the MCF-7 tumor cell line and their enhanced antioxidant and antimicrobial properties. International Journal of Nanomedicine, 2018; 13: 8013.
199. Cheloni G, Marti E, Slaveykova VI. Interactive effects of copper oxide nanoparticles and light to green alga *Chlamydomonas reinhardtii*. Aquatic Toxicology, 2016; 170: 120–128.
200. Bapat RA, Chaubal TV, Joshi CP, Bapat PR, Choudhury H, Pandey M, Gorain B, Kesharwani P. An overview of application of silver nanoparticles for biomaterials in dentistry. Materials Science and Engineering: C, 2018; 91: 881–898.
201. Gupta D, Singh A, Khan AU. Nanoparticles as efflux pump and biofilm inhibitor to rejuvenate bactericidal effect of conventional antibiotics. Nanoscale Research Letters, 2017; 12(1): 1–6.
202. Singh D, Rathod V, Ninganagouda S, Herimath J, Kulkarni P. Biosynthesis of silver nanoparticle by endophytic fungi *Pencillium* sp. isolated from *Curcuma longa* (turmeric) and its antibacterial activity against pathogenic gram negative bacteria. Journal of Pharmacy Research, 2013; 7(5): 448–453.
203. Gupta H. Role of nanocomposites in agriculture. Nano Hybrids and Composites, 2018; 20: 81–89.
204. Chauhan R, Kumar A, Abraham J. A biological approach to the synthesis of silver nanoparticles with *Streptomyces* sp JAR1 and its antimicrobial activity. Scientia Pharmaceutica, 2013; 81(2): 607–624.
205. Pareek V, Gupta R, Panwar J. Do physico-chemical properties of silver nanoparticles decide their interaction with biological media and bactericidal action? A review. Materials Science and Engineering: C, 2018; 90: 739–749.
206. Borisov VB, Siletsky SA, Nastasi MR, Forte E. ROS defense systems and terminal oxidases in bacteria. Antioxidants, 2021; 10(6): 839.

207. Vinutha J, Bhagat D, Bakthavatsalam N. Nanotechnology in the management of polyphagous pest *Helicoverpa armigera*. Journal of Academia and Industrial Research, 2013; 1(10): 606–608.

208. Khatami M, Iravani S, Varma RS, Mosazade F, Darroudi M, Borhani F. Cockroach wings-promoted safe and greener synthesis of silver nanoparticles and their insecticidal activity. Bioprocess and Biosystems Engineering, 2019; 42(12): 2007–2014.

209. Shukla G, Gaurav SS, Rani V, Singh A, Rani P, Verma P, Kumar B. Evaluation of larvicidal effect of mycogenic silver nanoparticles against white grubs (*Holotrichia* sp). Journal of Advanced Scientific Research, 2020; 11(1): 296–304.

210. Amin MA, Ismail MA, Badawy AA, Awad MA, Hamza MF, Awad MF, Fouda A. The potency of fungal-fabricated selenium nanoparticles to improve the growth performance of *Helianthus annuus* L. and control of cutworm *Agrotis ipsilon*. Catalysts, 2021; 11(12): 1551.

211. Arunthirumeni M, Veerammal V, Shivakumar MS. Biocontrol efficacy of mycosynthesized selenium nanoparticle using *Trichoderma* sp. on insect pest *Spodoptera litura*. Journal of Cluster Science, 2021; 1–9.

212. Jameel M, Shoeb M, Khan MT, Ullah R, Mobin M, Farooqi MK, Adnan SM. Enhanced insecticidal activity of thiamethoxam by zinc oxide nanoparticles: a novel nanotechnology approach for pest control. ACS Omega, 2020; 5(3): 1607–1615.

213. Ishwarya R, Vaseeharan B, Kalyani S, Banumathi B, Govindarajan M, Alharbi NS, Kadaikunnan S, Al-Anbr MN, Khaled JM, Benelli G. Facile green synthesis of zinc oxide nanoparticles using *Ulva lactuca* seaweed extract and evaluation of their photocatalytic, antibiofilm and insecticidal activity. Journal of Photochemistry and Photobiology B: Biology, 2018; 178: 249–258.

214. Hameed RS, Fayyad RJ, Nuaman RS, Hamdan NT, Maliki SA. Synthesis and characterization of a novel titanium nanoparticles using banana peel extract and investigate its antibacterial and insecticidal activity. Journal of Pure and Applied Microbiology, 2019; 13(4): 2241–2249.

215. Ali FAA, Alam J, Shukla AK, Alhoshan M, Khaled JM, Al-Masry WA, Alharbi NS, Alam M. Graphene oxide-silver nanosheet-incorporated polyamide thin-film composite membranes for antifouling and antibacterial action against *Escherichia coli* and bovine serum albumin. Journal of Industrial and Engineering Chemistry, 2019; 80: 227–238.

216. Nguyen N-YT, Grelling N, Wetteland CL, Rosario R, Liu H. Antimicrobial activities and mechanisms of magnesium oxide nanoparticles (nMgO) against pathogenic bacteria, yeasts, and biofilms. Scientific Reports, 2018; 8(1): 1–23.

217. Mao B-H, Chen Z-Y, Wang Y-J, Yan S-J. Silver nanoparticles have lethal and sublethal adverse effects on development and longevity by inducing ROS-mediated stress responses. Scientific Reports, 2018; 8(1): 1–16.

218. Benelli G. Green synthesized nanoparticles in the fight against mosquito-borne diseases and cancer – a brief review. Enzyme and Microbial Technology, 2016; 95: 58–68.

219. Benelli G. Plant-mediated synthesis of nanoparticles: a newer and safer tool against mosquito-borne diseases? Asian Pacific Journal of Tropical Biomedicine, 2016; 6(4): 353–354.

220. Jiang X, Miclăuş T, Wang L, Foldbjerg R, Sutherland DS, Autrup H, Chen C, Beer C. Fast intracellular dissolution and persistent cellular uptake of silver nanoparticles in CHO-K1 cells: implication for cytotoxicity. Nanotoxicology, 2015; 9(2): 181–189.

221. Shang Y, Hasan M, Ahammed GJ, Li M, Yin H, Zhou J. Applications of nanotechnology in plant growth and crop protection: a review. Molecules, 2019; 24(14): 2558.

222. Priyanka P, Kumar D, Yadav A, Yadav K. Nanobiotechnology and its application in agriculture and food production. *In*: Nanotechnology for Food, Agriculture, and Environment. 2020; Springer, Cham. pp. 105–134.

223. Naderi M, Danesh-Shahraki A. Nanofertilizers and their roles in sustainable agriculture. International Journal of Agriculture and Crop Sciences (IJACS), 2013; 5(19): 2229–2232.

224. Elemike EE, Uzoh IM, Onwudiwe DC, Babalola OO. The role of nanotechnology in the fortification of plant nutrients and improvement of crop production. Applied Sciences, 2019; 9(3): 499.

225. Zulfiqar F, Navarro M, Ashraf M, Akram NA, Munné-Bosch S. Nanofertilizer use for sustainable agriculture: advantages and limitations. Plant Science, 2019; 289: 110270.

226. Gupta M, Gupta S. An overview of selenium uptake, metabolism, and toxicity in plants. Frontiers in Plant Science, 2017; 7: 2074.

227. Bai H, Zhang Z, Guo Y, Yang G. Biosynthesis of cadmium sulfide nanoparticles by photosynthetic bacteria *Rhodopseudomonas palustris*. Colloids and Surfaces B: Biointerfaces, 2009; 70(1): 142–146.

228. Khan Z, Nisar MA, Hussain SZ, Arshad MN, Rehman A. Cadmium resistance mechanism in *Escherichia coli* P4 and its potential use to bioremediate environmental cadmium. Applied Microbiology and Biotechnology, 2015; 99(24): 10745–10757.

229. Kour D, Rana KL, Yadav N, Yadav AN, Singh J, Rastegari AA, Saxena AK. Agriculturally and industrially important fungi: current developments and potential biotechnological applications. *In*: Recent Advancement in White Biotechnology through Fungi. 2019; Springer, Cham. pp. 1–64.

230. Xie Y, Cheng W, Tsang PE, Fang Z. Remediation and phytotoxicity of decabromodiphenyl ether contaminated soil by zero valent iron nanoparticles immobilized in mesoporous silica microspheres. Journal of Environmental Management, 2016; 166: 478–483.

231. Yadav AN, Mishra S, Singh S, Gupta A. Recent Advancement in White Biotechnology through Fungi. Vol. 1: Diversity and Enzymes Perspectives. 2019; Springer, Cham. DOI: 10.1007/978-3-030-10480-1.

232. Yadav AN, Mishra S, Kour D, Yadav N, Kumar A. Agriculturally Important Fungi for Sustainable Agriculture. Volume 2: Functional Annotation for Crop Protection. 2020; Springer, Cham. DOI: 10.1007/978-3-030-48474-3.

233. Chen J, Lin Z, Ma X, Evidence of the production of silver nanoparticles via pretreatment of *Phoma* sp. 3.2883 with silver nitrate. Letters in Applied Mmicrobiology, 2003; 37(2): 105–108.

234. Purnomo AS, Putra SR, Shimizu K, Kondo R. Biodegradation of heptachlor and heptachlor epoxide-contaminated soils by white-rot fungal inocula. Environmental Science and Pollution Research, 2014; 21(19): 11305–11312.

235. Gadd GM. Biosorption: critical review of scientific rationale, environmental importance and significance for pollution treatment. Journal of Chemical Technology & Biotechnology: International Research in Process, Environmental & Clean Technology, 2009; 84(1): 13–28.

236. Joshi B. Evaluation and characterization of heavy metal resistant fungi for their prospects in bioremediation. Journal of Environmental Research and Development, 2014; 8(4): 876–882.

237. Dixit R, Malaviya D, Pandiyan K, Singh UB, Sahu A, Shukla R, Singh BP, Rai JP, Sharma PK, Lade H. Bioremediation of heavy metals from soil and aquatic environment: an overview of principles and criteria of fundamental processes. Sustainability, 2015; 7(2): 2189–2212.

238. Saxena P, Bhattacharyya AK, Mathur N. Nickel tolerance and accumulation by fila-
      mentous fungi from sludge of metal finishing industry. Geomicrobiology Journal,
      2006; 23(5): 333–340.
239. Ezzouhri L, Castro E, Moya M, Espinola F, Lairini K. Heavy metal tolerance of fila-
      mentous fungi isolated from polluted sites in Tangier, Morocco. African Journal of
      Microbiology Research, 2009; 3(2): 35–48.
240. Xu X, Xia L, Huang Q, Gu-DJ, Chen W. Biosorption of cadmium by a metal-
      resistant filamentous fungus isolated from chicken manure compost. Environmental
      Technology, 2012; 33(14): 1661–1670.
241. Bhargavi S, Savitha J. Arsenate resistant *Penicillium coffeae*: a potential fungus for
      soil bioremediation. Bulletin of Environmental Contamination and Toxicology, 2014;
      92(3): 369–373.
242. Maheswari S, Murugesan A. Remediation of arsenic in soil by *Aspergillus nidulans*
      isolated from an arsenic-contaminated site. Environmental Technology, 2009;
      30(9): 921–926.

9 Nanotechnology in Biotic Stress Management

Future Challenges and Opportunities

Ashish Kumar Singh, Krishna Kant Mishra, Jeevan Bettanayaka, Amit Umesh Paschapur, Gaurav Verma, and Lakshmi Kant

CONTENTS

9.1 INTRODUCTION

In their physical environment plants are surrounded by a complex set of environmental stresses that create adverse effects on their growth and development. Biotic stresses induced by microbes are considered to be one of the major constraints on crop yield losses and poor-quality produce [1]. Around 20–40% decline in agricultural productivity has been estimated globally due to various diseases caused by phytopathogenic fungi, bacteria, viruses, and nematodes [2]. The current world population of 7.6 billion is projected to reach around 9.8 billion in 2050. Roughly 800 million people in the world may remain hungry and the proportion of underfed people may be expected to reach >9% by 2030, thus it will be a challenge to fulfill the food demand of the growing population. Current disease management progress and strategies are not enough to mitigate the food crisis by 2050. Although the green revolution has brought significant changes to crop yield, at present crop production remains stagnant. Demand for food has been dramatically increasing, so now a second green revolution is needed, along with strategic research reforms to increase agricultural production and bring sustainability. Although chemical pesticides have

DOI: 10.1201/9781003322122-9

been able to contain insect pests and diseases, they have certain limitations, such as toxicity to the environment, expense, lack of effect when pest/disease pressure is high and possible contribution to pathogen resistance with their excessive use [3]. Therefore, it is essential to search for alternative efficient and safe agents to mitigate the yield losses caused by biotic stresses, with a view to making agriculture more sustainable [4–6].

Nanotechnology is an emerging interdisciplinary field. It has attracted significant attention in the agriculture sector by reducing yield losses as well as the excessive use of toxic chemicals. Nano denotes "one billionth" and deals with extremely small things, which are about 1–100 nm [1]. Nanoparticles are expected to deal with the limitations of existing conventional strategies. Nano-pesticides have unique physiological characteristics, a high surface area-to-volume ratio, small sizes, thus improving their reactivity and biochemical activity, and advanced formulations, which can increase their solubility and penetration within plants as well as the insect body [7, 8]. Recent research progress and the role of various nanoparticles in the management of insect pests and diseases and their impact on plant resistance are discussed below.

9.2   NANOMATERIALS IN PLANT DISEASE MANAGEMENT

Nanomaterials have been reported to enhance resistance in crop plants against various types of diseases caused by plant pathogenic fungi, bacteria, and nematodes (Table 9.1). Nanomaterials also have antibacterial and antifungal activity. The inhibitory action of nanoparticles on fungi and bacteria includes disrupting cell walls, enhancing the permeability of the cytoplasmic membrane, disrupting the bacterial envelope, direct attachment to the cell surface, DNA damage, cell cycle arrest, inhibition of the normal budding process due to the destruction of the membrane integrity,

TABLE 9.1
Nanoparticles (NPs) and Their Effects on Plant Growth

Nanoparticles	Plant	Mechanism	References
ZnO NPs	Wheat (*Triticum aestivum*)	Enhance seed germination	[17]
SiO_2 and TiO_2 NPs	Rice	Improve plant growth attributes	[18]
TiO_2 and ZnO NPs	Beetroot	Improve plant growth and shoot dry matter	[19]
TiO_2 NPs	*Brassica napus*	Positive effect on germination and seedling vigor	[20]
SiO_2 NPs	*Larix olgensis*	Improve plant height, root length and chlorophyll content	[21]
MgO NPs	Tomato	Induce disease resistance	[22]
Chitosan NPs	Apples	Reduce microbial growth	[23]
SiO_2 and Mo NPs	Rice	Regulate seed germination	[24]
$Cu(OH)_2$	*Lactuca sativa*	Activate antioxidant system	[25]
FeS_2 NP	Spinach	Improve plant growth	[26]

and inhibition of enzyme activity and reactive oxygen species (ROS) generation, and this finally leads to death [9, 10].

A wide range of nano-formulations such as silver nanoparticles (AgNPs), Cu composites (CuO NPs), ZnO nanoparticles (ZnO NPs), chitosan nanoparticles, titanium dioxide (TiO$_2$ NPs), (Table 9.2), are potential candidates that are in use in agricultural systems, including crop protection (Figure 9.1) [11]. Silver nanoparticles were reported to inhibit many fungal pathogens, including *Alternaria alternata, A. brassicicola, Botrytis cinereal, Fusarium oxysporum* f. sp. *lycopersici* etc. [12]. Similarly, Ag nanoparticles also showed antibacterial activity against *Erwinia* sp., *Bacillus* sp.,

TABLE 9.2
Nanoparticles (NPs) and Their Antimicrobial Activity

Nanoparticles	Pathogen	Mechanism	References
AgNPs	*Alternaria alternata, A. brassicicola, A. solani, Botrytis cinerea, F. oxysporum f.* sp. *lycopersici, Pythium aphanidermatum, Stemphylium lycopersici, Bipolaris sorokiniana* and *Magnaporthe grisea*	Antifungal activity	[27, 12]
AgNPs	*Pseudomonas aeruginosa, Escherichia coli, Bacillus subtilis*	Antibacterial activity	[28]
AgNPs and Cs-Ag nanocomposite	*Pseudomonas syringae* pv. *syringae*	Antibacterial activity	[29]
Chitosan NPs	*Fusarium solani, Aspergillus niger*	Antifungal activity	[30]
AuNPs	*Puccinia graminis tritici, Aspergillus flavus, Aspergillus niger* and *Candida albicans*	Antifungal activity	[31]
Nano Si-Ag	*Pythium ultimum, Magnaporthe grisea, Colletotrichum gloeosporioides, Botrytis cineria, Rhizoctonia solani, Pseudomonas syringae, Xanthomonas compestris* pv. *vesicatoria*	Both antifungal and antibacterial activity	[32]
ZnO NPs	*Alternaria alternata, Botrytis cinerea, Aspergillus niger, Fusarium oxysporum* and *Penicillium expansum*	Antifungal activity	[16]
TiO$_2$ NPs	*Fusarium solani*	Inhibit fungal growth	[33]
TiO$_2$ NPs	*Xanthomonas hortorum* pv. *pelargonii, X. axonopodis* pv. *Poinsettiicola*	Antibacterial activity	[34]
AuNPs	*Escherichia coli, Staphylococcus aureus*	Antibacterial activity	[35]

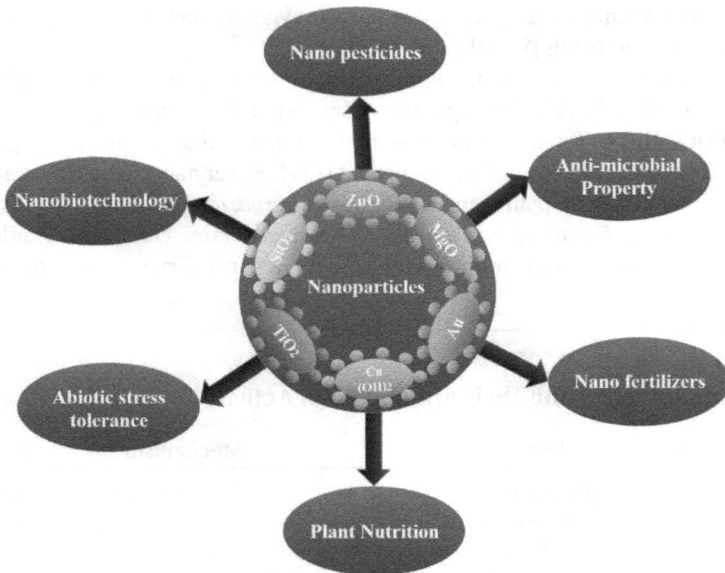

FIGURE 9.1   Applications of nanoparticles in various fields of agriculture.

Pseudomonas sp., *Staphylococcus* sp., etc. [13, 14]. CuO NPs showed antifungal activity against *Botrytis cinerea*, *Colletotrichum graminicola*, *Rhizoctonia solani*, *Colletotrichum musae*, *Magnaporthe oryzae*, *Penicillium digitatum*, and *Sclerotium rolfsii* [15]. ZnO NPs found promising results against *Alternaria alternata*, *Botrytis cinerea*, *Aspergillus niger*, *Fusarium oxysporum* and *Penicillium expansum* fungi [16].

9.3   NANO-PESTICIDES FOR INSECT PEST MANAGEMENT

Insect pests are major biotic stress-causing agents that are reported to cause 18–20% crop losses worldwide [36, 37]. To manage the insect pest menace, several pest management tactics, such as cultural, mechanical, biological and chemical methods, are followed by farmers. Among them, the chemical control method in which synthetic pesticides are used has become popular. It is now considered as an important strategy for enhancing crop productivity and achieving satisfactory crop protection in modern agriculture [38, 39]. The conventional group of insecticides used for the management of insect pests in traditional pest management practices has several major disadvantages, like high dosage per unit crop, drift hazards, operational hazards and residues in environment, plants and in marketable produce. They also affect non-target vegetation and organisms. So, they need to be replaced by an alternative pest control strategy that can overcome the above lacunas. Nano-pesticides are one of the alternatives to overcome the lacunas of the conventional group of insecticides [39, 40]. The major benefits of these nanoparticles include the improved solubility of active ingredients, better stability of formulation, slow release of active ingredient and improvement in mobility caused by smaller particle size and higher surface area.

The mode of action against target pests is expected to be enhanced with nanoparticles, as opposed to bulk materials. Moreover, nano-formulations provide systemic properties, uniform leaf coverage and improved soil properties to support their constructive use in agriculture [41–43].

9.4 FORMULATIONS OF NANO-PESTICIDES

Research in nanotechnology has led to the development of nano-formulations of insecticides that can be applied in crop protection. They include nano-insecticides, nano-herbicides, nano-fungicides and nano-nematicides. Nano-insecticides are formulated according to their intended purpose as formulations that improve solubility, release active ingredients slowly and prevent degradation [44]. To achieve these purposes, the chemical nature of the carrier molecule has been modified and classified as organic polymer-based and lipid-based formulations, nano-sized metals and metal oxides and clay-based nanomaterials [39]. A few important nano-insecticide formulations are presented in Figure 9.2.

Recently the promising results obtained from the use of nanoparticles for insect pest management have opened up a new field of research in integrated pest management. Some of the synthetic nano-insecticide formulations and nano-biocides, along with their active ingredients used against several target insect pests, are mentioned in Tables 9.3 and 9.4, respectively.

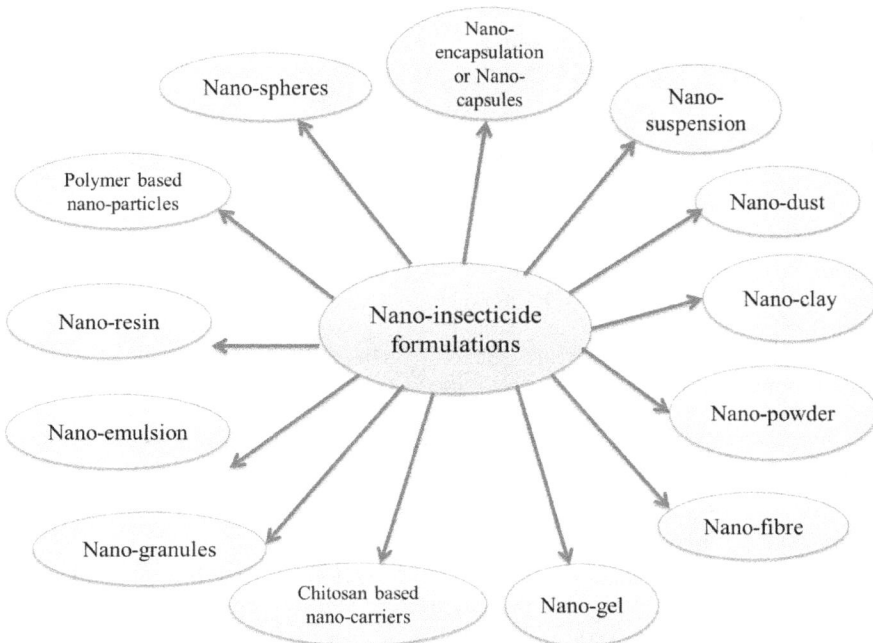

FIGURE 9.2 Nano-insecticide formulations.

TABLE 9.3
Synthetic Nano-Insecticides Used Against Target Insect Pests

Nanoparticle	Active ingredient	Target organism	References
Nano-encapsulation	Imidacloprid	*Martianus dermestoides, Culex quinquefasciatus*	[45–47]
Nano-capsules	Avermectin	*Martianus dermestoides*	[46]
Nano-capsules	β-Cyfluthrin	*Aedes aegypti* and sucking pests	[48, 49]
Nano-capsules	Etofenprox	Lepidopteran insects	[50]
Nano-capsules	Deltamethrin	Coleopterans and lepidopterans	[51]
Nano-capsules	Permethrin	*Culex quinquefasciatus*	[52]
Nano-capsules	Carbaryl	Soil arthropods	[53]
Nano-clay	Cyromazine	Insect growth regulator	[54]
Nano-fiber	Pheromones	Pink boll worm and fruit flies	[55]
Nano-gel	Cypermethrin	Termites and lepidopteran larvae	[56]
Nano-gel	Aldicarb	Ovicidal against coleopterans and lepidopterans	[57]
Nano-granules	Imidacloprid or cyromazine	Veterinary insect pests and sucking pests	[58]
Nano-micelle	Carbofuran	Soil arthropods	[59]
Nano-emulsion	Chlorpyrifos	Sucking pests and defoliators	[60]
Nano-resin	Pheromones	Cotton boll worms	[61]
Nano-spheres	Carbaryl	Coleopteran and lepidopteran defoliators	[62]
Nano-suspension	Carbofuran	Soil arthropods	[63]
Nano-powder	Novaluron	Insect growth regulators	[64]
Nano-DEPA (diethylphenylacetamide)	DEPA mosquito	*Culex quinquefasciatus*	[65]
Chitosan-coated nano-formulations	Pyrifluquinazon	Green peach aphid (*Myzus persicae*)	[66]

Apart from the above-mentioned synthetic insecticide, plant and microbial-based nano-formulations, there are several metal-based nano-insecticides that have found a prominent place in insect pest management (Table 9.5).

Nanotechnology offers a wonderful tool for developing novel formulations of eco-friendly insecticides. Most nano-insecticide formulations are highly target-specific, due to their targeted delivery and controlled release, and this can improve insecticide utilization and reduce residue and pollution. For example, nano-microcapsule

TABLE 9.4
Nano-Biocides Used Against Target Insect Pests

Nanoparticle	Active ingredient	Target organism	References
	Plant-based nano-bioinsecticides		
Nano-encapsulation	Essential oil of *Carum copticum*	Diamondback moth (*Plutella xylostella*)	[67]
Polyethylene glycol-coated nanoparticles	Essential oil of garlic	Red flour beetle (*Tribolium castaneum*)	[68]
Nanoparticles	Azadirachtin	Insect-feeding deterrents	[69]
Nano-suspensions	*Jatropha gossypifolia, Euphorbia tirucalli, Pedilanthus tithymaloides, Alstonia macrophylla*	*Aedes aegypti*	[70]
Nanoparticles	*Nelumbo nucifera*	*Anopheles stephensi, Culex quinquefasciatus*	[71]
Nanoparticles	*Moringa oleifera* seeds	*Stegomya aegypti*	[72]
Nanoparticles	*Artemisia arborescens* (oil)	*Bemisia tabaci*	[73]
Nano-emulsions	*Mukia maderaspatana*	*Culex quinquefasciatus, Aedes aegypti*	[74]
Nano-suspensions	*Plumeria rubra*	*Aedes aegypti, Anopheles stephensi*	[75]
	Microbial-based nano-bioinsecticides		
Chitosan nano-carrier	*Nomuraea rileyi*	Tobacco cutworm (*Spodoptera litura*)	[76]
Nano-encapsulation	*Aspergillus niger*	*Anopheles stephensi, Culex quinquefasciatus, Aedes aegypti*	[77]
Nanoparticles	*Chrysosporium tropicum*	*Anopheles stephensi, Culex quinquefasciatus*	[78]
Nanoparticles	*Bacillus sphaericus*	*Culex quinquefasciatus*	[79]

formulations have slow release and protection performance because they have been prepared using light-, heat-, humidity-, enzyme- and soil pH-sensitive high-polymer materials to deliver pesticides. Nano-pesticide formulations improve adhesion of droplets on the plant surface (reducing drift losses) which in turn improves the dispersion and bio-activity of the active ingredient of pesticide molecules. Therefore, nano-pesticides will have high efficacy compared to conventional pesticide formulations

TABLE 9.5
Metal-Based Nano-Formulations Used Against Target Insect Pests

Nanoparticle	Active ingredient	Target organism	References
Metal-based nano-bioinsecticides			
Nano-encapsulation	Nano-silica	*Anopheles stephensi, Culex quinquefasciatus, Aedes aegypti*	[80]
Nano-encapsulation	Silver, aluminum oxide, zinc oxide and titanium dioxide nanoparticles	Several insect pests and pathogens	[81]
Nano dust	Nano-Al_2O_3 dust	Rice weevil (*Sitophilus oryzae*), *Rhyzopertha dominica* (F.)	[82]
Nano-encapsulation	Silver and gold nanoparticles	*Aphis nerii Spodoptera litura*	[83, 84]
Nanoparticles	Nano-cadmium sulfide (CdS)	*Spodoptera litura*, fall army warm	[81]

and, due to their small size, improvable pesticide droplet ductility, wettability and target adsorption have made these nano-pesticides an efficient and environmentally friendly technology for insect pest management. Moreover, nanoparticle-mediated gene transfer may also play an important role in developing new insect-resistant varieties [85]. Nano-pesticides are an extraordinary means of setting up an eco-friendly and sustainable agriculture system because they reduce overall chemical usage, decrease toxic residues, halt insecticide resistance and insect resurgence and enhance overall crop protection [43].

Before embarking on the use of nano-insecticides on a commercial scale for insect pest management, the risk that nanoparticles (nano-pesticides) may pose to human and environment health has to be closely studied, because nano-pesticides are novel molecules that may create new kinds of contamination of soils and waterways, since they are apparently much more persistent and have higher degrees of toxicity when compared to their traditional counterparts. Therefore, a better understanding of the fate and effect of nano-pesticides after their application is required. It is a good thing that all necessary safety precautions are taken before the decision is made to go ahead and use new technologies on a large scale [86].

9.5 NANOTECHNOLOGY FOR NEMATODE MANAGEMENT

Nematodes are the most abundant metazoans on this planet. Categorically, free-living nematodes and parasites play an important role in the ecosystem. Free-living nematodes carry out crucial mineralization of nutrients and some serve as an important model organism in unraveling and understanding biology. This parasite group includes nematodes causing disease in animals, humans and plants. Plant parasitic nematodes (PPNs) are the major biotic stress of crops and yield constraint remains hidden, but incurred loss of 173 billion dollar globally [87]. In India this loss was estimated to be around Rs102,039.79 million (1.58 billion USD) annually [88]. In developing countries, the damage caused by nematodes is highly underestimated due to their microscopic size, soil-borne habitat and a lack of awareness about nematodes. But, in the context of climate change, growth in protected cultivation, withdrawals of nematicides and inefficient management practices may increase the chances of nematode-induced severity and dispersal in new areas [89]. Therefore, to ensure global food security, an estimation of nematode population in the soil, crop-wise distribution pattern of key nematode pests, diversity anddetermination of damage potential are key points in protecting plant health from nematode assault and their dispersal. More than 4100 PPN species are impacting global agricultural crops in a serious way [90]. The *Meloidogyne* genus belongs to the group of root knot nematode (RKN) that contains more than 90 species. RKN (*Meloidogyne incognita*) under protected cultivation remains a major challenge to growers in India. Recently, other RKNs, such as *Meloidogyne graminicola* in the rice-wheat cropping system and *Meloidogyne enterolobii* in guava, have emerged as serious threats. The most-dreaded PPNs like potato cyst nematode (*Globodera* spp.) are serious concerns for India, Africa and Portugal, due to their long-term survival strategy of eggs under hard cysts and dependence on the potato root diffusates to continue their life cycle. Molya disease of wheat, caused by *Heterodera avenae* in Rajasthan and Haryana, is imposing significant yield losses. The losses incurred by nematodes are not limited to quantitative yield losses; they also affect the quality of potato tubers, ginger rhizomes and turmeric and carrot roots and predispose to blast diseases as a result of the infection of soil-borne pathogens, vector of viruses and enhanced suscepti-bility of rice plants, because of the below-ground attack of *Meloidogyne graminicola* [91]. Several nematode management approaches have been investigated and have demonstrated varying effects. Conventional practices, such as use of cover crop [92], application of soil solarization, soil amendments through green manuring, enforce-ment and regulation of plant quarantine, utilization of biocontrol agents (bacteria, fungi), including nematode antagonists and plant growth-promoting rhizobacteria are the main options for nematode control. Deployment of resistance genotypes in breeding programs are the most eco-friendly and sustainable approach but limited resistance source, broad host range of PPNs, climate vulnerability, resistance break-down, high selection pressure and emergence of virulent population are major factors in reducing effectiveness at field level. Application of chemical nematicides is the most effective and time-saving management strategy against nematodes. However, with the recent ban and withdrawals of several chemical nematicides such as methyl bromide and aldicarb, newer replacement nematicides (fluopyram, fluensulfone) have been developed. An innovative approach in targeting nematodes through host-delivered RNAi mechanism is novel and effective; so far, the efficiency of this tech-nique has proven promising under laboratory conditions but it has yet to be delivered a farmer's field. Henceforth, the development of improved specific methodology and science-led technique that is easy in practice would be the best strategy against nematodes.

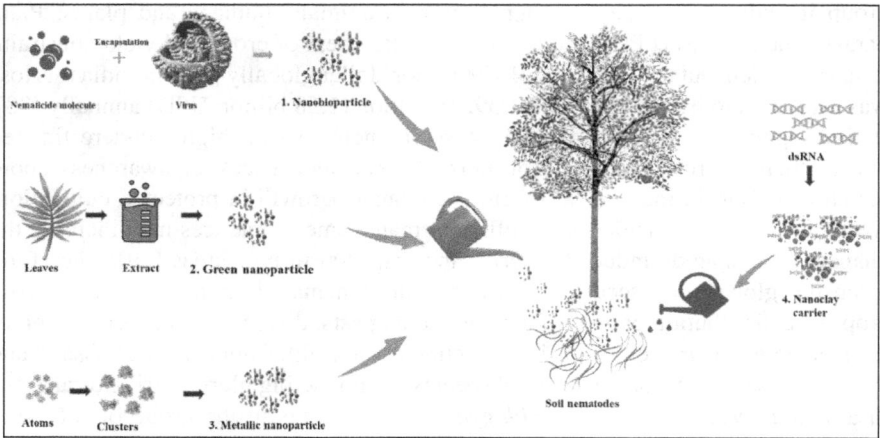

FIGURE 9.3 Approaches of nanotechnology for nematode management.

Nanotechnology is a fast-developing science with broad applicability. It deals with the creation, synthesis and application of any material attenuated to the scale of less than 100 nm exhibiting novel properties, which differs from parental material due to nanoscale size and greater surface-to-volume proportions. The development of nano-formulations including nano-bioparticles, green nanoparticles, metal nanoparticles and delivery of dsRNA with nano clay encapsulation to target nematode pests (Figure 9.3) is the major alternative solution [93]. Nanoparticles are the basic building blocks and come in different shapes and chemical properties. They offer good mobility, reactivity and biological activity into the targeted cellular system [94]. These nanoparticles could be comprised of many materials or a single one, including metals, semiconductors, metal oxides, polymers, clay, carbon and biological peptides.

A nanobioparticle is a nano-organism (virus) encapsulated with nano secondary metabolites obtained from antagonistic microbes or nematicidal active ingredients. These nano-organisms serve as carriers or delivery vehicles of nematicides by invading a targeted root and transporting nanoparticles of interest to the tissue system. offering high target specificity, wider applicability at small dosage and enhanced mobility of nematicides to cover broader root zone areas. To manage the endoparasitic nematodes of corn, pesticide was encapsulated into a tobacco mild green mosaic virus to be delivered at target for enhanced diffusion. This encapsulated virus particle enhanced the diffusion of pesticides to the root level where nematode interacts and enhanced nematode management efficacy [95]. Nematicidal molecules derived from secondary metabolites of *Streptomyces avermitilis* are very potent nematicides but their mobility in soil is poor, hence they provide nematode protection in limited zone of soil. Engineering microbes with endotoxin-coding genes is a new prospect in the management of nematodes. Frugivorous pine wood nematode (*Bursaphelenchus xylophilus*) is a destructive pest of pine which is of importance as regards quarantine. For their management, the *Bacillus thuringiensis* gene (*cry5Ba3Φ*) was transformed in the fungus on which pine wood nematode feeds to complete its life cycle. When nematods fed on these transformed fungi, their fitness decreased [96]. Flash nanoprecipitation has been utilized to produce and amplify abamectin-loading capacity and encapsulation efficiency with amphiphilic block copolymer nanoparticles to target *M. incognita* [97].

Green nanoparticles have been synthesized using plant extracts to control nematodes in soil. Plant-based green nanoparticles offer advantages for more stability, diversity of shape, easy rate of synthesis process, eco-friendliness and low environmental toxicity. Green silver nanoparticle synthesized from algal extract successfully reduced RKN infestation in eggplant [98]. Silver nanoparticles synthesized from dog nettle grass (*Urtica urens*) were found to be effective for the management of *M. incognita* in comparison to normal silver nanoparticles and petroleum ether [99]. Similarly, silver nanoparticles have been extracted from leaves of *Conyza dioscoridis*, *Melia azedarach*, and *Moringa oleifera*. Among them, *Conyza dioscoridis* showed great nematicidal potential against RKN [100]. Many other plants have been used for the synthesis of green nanoparticles, such as *Sesbania drummondii*, *Clethra barbinervis*, *Brassica juncea*, *Ilex crenata*, *Maytenus fournieri* and *Acanthopanax sciado phylloides* [101].

Direct use of metallic nanoparticles (silver and sulfur-based) has shown promising potential in nematode mortality, and hence may be used as alternative nematode management strategy in future. Silver nanoparticles have shown nematicidal potential by inducing cellular oxidative stress [102] and pave the way to find novel nematicidal nanoparticles. Nanoparticle Al_2O_3 demonstrated nematicidal toxicity by reducing reproduction, and direct inhibition in growth and development of nematodes [103, 104]. The efficacy of silica nanoparticles investigated on *Caenorhabditis elegans* showed significant reproductive organ degeneration [105]. Furthermore, the toxicity of various nanoparticles (Fe_2O_3, titanium oxide, ZnO, silver and Al_2O_3) against *C. elegans* has been assayed [103, 106]. Against PPNs the application of silver nanoparticles has shown reduced nematode population and increase in turf grass quality [107]. Nematotoxicity of silver, titanium oxide (TiO_2) and silicon oxide against juveniles of RKN (*M. incognita*) was assayed in the laboratory and showed nematicidal action in the treatment of silver and TiO_2 [108]. Treatment of *M. incognita* with AgNP at 30–150 mg/mL inactivated 99% of juveniles. Treatment of turf grass with AgNP at 150 mg/mL significantly reduced *M. graminis* population up to 92%. Similarly, toxicity assays in lab and pot conditions of gold nanoparticles, silicon oxide, silver and titanium oxide against RKN have shown promising results in reducing nematode damage [108–110]. The mode of action of these nanoparticles has not been specified but it has been reported to be disrupting cellular mechanism, membrane permeability, synthesis of ATP and oxidative stress [102, 106, 111]. The mode of action of nanoparticles is diverse and includes the inhibition of protein synthesis in targeted cells, DNA function and cell burst and affects the function of respiratory enzymes of targeted pest.

In the recent past, uptake of dsRNA has been validated and demonstrated through exogenous application to silence critical genes of phytopathogens, including virus, fungi and insects [112]. Since nematodes are feeders of root tissues the possibility of dsRNA spray treatment on plant roots and application through root dip, irrigation and on-site spray against nematode attack could be attractive avenues for the development of a novel strategy to protect crops from nematodes. Limited dsRNA uptake through roots is the major limitation against nematode attack. Development of BioClay to deliver dsRNA [113] offers hope to adapt and develp a strategy against nematodes. Abamectin encapsulated into red clover necrotic mosaic virus enhanced the zone of root protection against *Meloidogyne hapla* in tomato. Plants treated with this delivery system (virus encapsulated abamectin) provided increased efficacy of nematode control [114].

9.6   CONCLUSION AND FUTURE PERSPECTIVES

Future adaptation of nanotechnology may have a significant impact on plant disease management. The development of nano-formulations is being considered as eco-friendly alternatives for nematode management. Limited studies and information have been conducted with respect to applicability in pest and disease management. The high lethal toxicity of nanomaterials and their environmental stability are risks to the food chain. Therefore, identification and synthesis of nanoparticles that are specifically toxic for target pests, that are less stable in the environment and eco-friendly in nature are required. The preparation of nanoparticles from secondary metabolites obtained from microbes may offer a safe, potent and eco-friendly approach for pest and disease management.

REFERENCES

1. Khan AU, Khan M, Khan MM. Antifungal and antibacterial assay by silver nanoparticles synthesized from aqueous leaf extract of *Trigonella foenum-graecum*. Bio Nano Science, 2019; 9(3): 597–602.
2. Cabral-Pinto M, Inácio M, Neves O, Almeida AA, Pinto E, Oliveiros B, Ferreira da Silva EA. Human health risk assessment due to agricultural activities and crop consumption in the surroundings of an industrial area. Exposure and Health, 2020; 12(4): 629–640.
3. Jeevan B, Gogoi R, Sharma D, Manjunatha C, Rajashekara H, Ram D, Mishra KK, Mallikarjuna MG. Genetic analysis of maydis leaf blight resistance in subtropical maize (*Zea mays* L.) germplasm. Journal of Genetics, 2020; 99(1): 1–9.
4. Arora S, Sharma P, Kumar S, Nayan R, Khanna PK, Zaidi MG. Gold-nanoparticle induced enhancement in growth and seed yield of *Brassica juncea*. Plant Growth Regulation, 2012; 66(3): 303–310.
5. Usman M, Farooq M, Wakeel A, Nawaz A, Cheema SA, Rehman H, Ashraf I, Sanaullah M. Nanotechnology in agriculture: current status, challenges and future opportunities. Science of the Total Environment, 2020; 721: 137778.
6. Singh A, Tiwari S, Pandey J, Lata C, Singh IK. Role of nanoparticles in crop improvement and abiotic stress management. Journal of Biotechnology, 2021; 337: 57–70.
7. Salajegheh M, Yavarzadeh M, Payandeh A, Akbarian MM. Effects of titanium and silicon nanoparticles on antioxidant enzymes activity and some biochemical properties of *Cuminum cyminum* L. under drought stress. Banat's Journal of Biotechnology, 2020; 11(21): 19–25.
8. Deka B, Babu A, Baruah C, Barthakur M. Nanopesticides: a systematic review of their prospects with special reference to tea pest management. Frontiers in Nutrition, 2021; 8: 686131. doi: 10.3389/fnut.2021.686131

9. Khorrami S, Zarrabi A, Khaleghi M, Danaei M, Mozafari MR. Selective cytotoxicity of green synthesized silver nanoparticles against the MCF-7 tumor cell line and their enhanced antioxidant and antimicrobial properties. International Journal of Nanomedicine, 2018; 13: 8013.

10. Yin IX, Zhang J, Zhao IS, Mei ML, Li Q, Chu CH. The antibacterial mechanism of silver nanoparticles and its application in dentistry. International Journal of Nanomedicine, 2020; 15: 2555.

11. Khan M, Khan AU, Hasan MA, Yadav KK, Pinto MM, Malik N, Yadav VK, Khan AH, Islam S, Sharma GK. Agro-nanotechnology as an emerging field: a novel sustainable approach for improving plant growth by reducing biotic stress. Applied Sciences, 2021; 11(5): 2282.

12. Kim SW, Jung JH, Lamsal K, Kim YS, Min JS, Lee YS. Antifungal effects of silver nanoparticles (AgNPs) against various plant pathogenic fungi. Mycobiology, 2012; 4: 53–58.

13. Shehzad A, Qureshi M, Jabeen S, Ahmad R, Alabdalall AH, Aljafary MA, Al-Suhaimi E. Synthesis, characterization and antibacterial activity of silver nanoparticles using *Rhazya stricta*. Peer J, 2018; 6: e6086.

14. Rodríguez-Serrano C, Guzmán-Moreno J, Ángeles-Chávez C, Rodríguez-González V, Ortega-Sigala JJ, Ramírez-Santoyo RM, Vidales-Rodríguez LE. Biosynthesis of silver nanoparticles by *Fusarium scirpi* and its potential as antimicrobial agent against uropathogenic *Escherichia coli* biofilms. Plos One, 2020; 15(3): e0230275.

15. Huang S, Wang L, Liu L, Hou Y, Li L. Nanotechnology in agriculture, livestock, and aquaculture in China. A review. Agronomy for Sustainable Development, 2015; 35(2): 369–400.

16. Jamdagni P, Khatri P, Rana JS. Green synthesis of zinc oxide nanoparticles using flower extract of *Nyctanthes arbor-tristis* and their antifungal activity. Journal of King Saud University-Science, 2018; 30(2): 168–175.

17. Awasthi A, Bansal S, Jangir LK, Awasthi G, Awasthi KK, Awasthi K. Effect of ZnO nanoparticles on germination of *Triticum aestivum* seeds. In: Macromolecular Symposia, 2nd International Conference on Soft Materials, Jaipur, India, 2017; 376 (1), p. 1700043.

18. Rizwan M, Ali S, Malik S, Adrees M, Qayyum MF, Alamri SA, Alyemeni MN, Ahmad P. Effect of foliar applications of silicon and titanium dioxide nanoparticles on growth, oxidative stress, and cadmium accumulation by rice (*Oryza sativa*). Acta Physiologiae Plantarum, 2019; 41(3): 1–2

19. Siddiqui ZA, Khan MR, Abd_Allah EF, Parveen A. Titanium dioxide and zinc oxide nanoparticles affect some bacterial diseases, and growth and physiological changes of beetroot. International Journal of Vegetable Science, 2019; 25(5): 409–430.

20. Mahmoodzadeh H, Nabavi M, Kashefi H. Effect of nanoscale titanium dioxide particles on the germination and growth of canola (*Brassica napus*). Journal of Ornamental Plants, 2013; 25–32.

21. Bao-Shan L, Chun-hui L, Li-jun F, Shu-chun Q, Min Y. Effect of TMS (nanostructured silicon dioxide) on growth of Changbai larch seedlings. Journal of Forestry Research, 2004; 15(2): 138–140.

22. Imada K, Sakai S, Kajihara H, Tanaka S, Ito S. Magnesium oxide nanoparticles induce systemic resistance in tomato against bacterial wilt disease. Plant Pathology, 2016; 65(4): 551–560.

23. Pilon L, Spricigo PC, Miranda M, de Moura MR, Assis OB, Mattoso LH, Ferreira MD. Chitosan nanoparticle coatings reduce microbial growth on fresh-cut apples while not affecting quality attributes. International Journal of Food Science & Technology, 2015; 50(2): 440–448.

24. Adhikari T, Kundu S, Rao AS. Impact of SiO_2 and Mo nano particles on seed germination of rice (*Oryza sativa* L.). International Jourmal of Agricultural Food Science Technology, 2013; 4(8): 809–816.

25. Zhao L, Ortiz C, Adeleye AS, Hu Q, Zhou H, Huang Y, Keller AA. Metabolomics to detect response of lettuce (*Lactuca sativa*) to Cu $(OH)_2$ nanopesticides: oxidative stress response and detoxification mechanisms. Environmental Science & Technology, 2016; 50(17): 9697–9707.

26. Srivastava G, Das CK, Das A, Singh SK, Roy M, Kim H, Sethy N, Kumar A, Sharma RK, Singh SK, Philip D. Seed treatment with iron pyrite (FeS_2) nanoparticles increases the production of spinach. RSC Advances, 2014; 4(102): 58495–58504.

27. Jo YK, Kim BH, Jung G. Antifungal activity of silver ions and nanoparticles on phytopathogenic fungi. Plant Disease, 2009; 93(10): 1037–1043.

28. Mohanta YK, Panda SK, Bastia AK, Mohanta TK. Biosynthesis of silver nanoparticles from *Protium serratum* and investigation of their potential impacts on food safety and control. Frontiers in Microbiology, 2017; 18(8): 626.

29. Shahryari F, Rabiei Z, Sadighian S. Antibacterial activity of synthesized silver nanoparticles by sumac aqueous extract and silver-chitosan nanocomposite against *Pseudomonas syringae* pv. *syringae*. Journal of Plant Pathology, 2020; 102(2): 469–475.

30. Xing K, Shen X, Zhu X, Ju X, Miao X, Tian J, Feng Z, Peng X, Jiang J, Qin S. Synthesis and in vitro antifungal efficacy of oleoyl-chitosan nanoparticles against plant pathogenic fungi. International Journal of Biological Macromolecules, 2016; 82(1):830–836.

31. Jayaseelan C, Ramkumar R, Rahuman AA, Perumal P. Green synthesis of gold nanoparticles using seed aqueous extract of *Abelmoschus esculentus* and its antifungal activity. Industrial Crops and Products, 2013; 45(1): 423–429.

32. Park HJ, Kim SH, Kim HJ, Choi SH. A new composition of nanosized silica-silver for control of various plant diseases. The Plant Pathology Journal, 2006; 22(3): 295–302.

33. Hamza A, Mohamed A, Hamed S. New trends for biological and non-biological control of tomato root rot, caused by *Fusarium solani*, under greenhouse conditions. Egyptian Journal of Biological Pest Control, 2016; 26(1): 89–96.

34. Norman DJ, Chen J. Effect of foliar application of titanium dioxide on bacterial blight of geranium and *Xanthomonas* leaf spot of poinsettia. HortScience, 2011; 46(3): 426–428.

35. Yuan Y, Liu F, Xue L, Wang H, Pan J, Cui Y, Chen H, Yuan L. Recyclable *Escherichia coli*-specific-killing AuNP–polymer (ESKAP) nanocomposites. ACS Applied Materials & Interfaces, 2016; 8(18): 11309–11317.

36. Arora R, Sandhu S. Insect–plant interrelationships. In: Breeding Insect Resistant Crops for Sustainable Agriculture. 2017; Springer, Singapore. pp. 1–44.

37. FAO. News article: New standards to curb the global spread of plant pests and diseases. 2021.

38. Zhang W. Global pesticide use: profile, trend, cost/benefit and more. Proceedings of the International Academy of Ecology and Environmental Sciences, 2018; 8(1): 1.

39. Zannat R, Rahman MM, Afroz M. Application of nanotechnology in insect pest management: a review. SAARC Journal of Agriculture, 2021; 19(2): 1–1.

40. Duhan JS, Kumar R, Kumar N, Kaur P, Nehra K, Duhan S. Nanotechnology: the new perspective in precision agriculture. Biotechnology Reports, 2017; 15: 11–23.

41. Kah M, Hofmann T. Nanopesticide research: current trends and future priorities. Environment International, 2014; 63: 224–235.

42. Bhan S, Mohan L, Srivastava CN. Nanopesticides: a recent novel ecofriendly approach in insect pest management. Journal of Entomological Research, 2018; 42(2): 263–270.

43. Rajna S, Paschapur A, Raghavendra K. Nanopesticides: its scope and utility in pest management. Indian Farmer, 2019; 6(1): 17–21.
44. Ragaei M, Sabry AK. Nanotechnology for insect pest control. International Journal of Science, Environment and Technology, 2014; 3(2): 528–545.
45. Flores-Céspedes F, Figueredo-Flores CI, Daza-Fernández I, Vidal-Peña F, Villafranca-Sánchez M, Fernández-Pérez M. Preparation and characterization of imidacloprid lignin–polyethylene glycol matrices coated with ethylcellulose. Journal of Agricultural and Food Chemistry, 2012; 60(4): 1042–1051.
46. Guan YF, Pearce RC, Melechko AV, Hensley DK, Simpson ML, Rack PD. Pulsed laser dewetting of nickel catalyst for carbon nanofiber growth. Nanotechnology, 2008; 19(23): 235604.
47. Bhan S, Mohan L, Srivastava CN. Relative larvicidal potentiality of nano-encapsulated temephos and imidacloprid against *Culex quinquefasciatus*. Journal of Asia-Pacific Entomology, 2014; 17(4): 787–791.
48. Kumar S, Dilbaghi N, Saharan R, Bhanjana G. Nanotechnology as emerging tool for enhancing solubility of poorly water-soluble drugs. Bionanoscience, 2012; 2(4): 227–250.
49. Loha KM, Shakil NA, Kumar J, Singh MK, Srivastava C. Bio-efficacy evaluation of nanoformulations of β-cyfluthrin against *Callosobruchus maculatus* (Coleoptera: Bruchidae). Journal of Environmental Science and Health, Part B, 2012; 47(7): 687–691.
50. Hwang IC, Kim TH, Bang SH, Kim KS, Kwon HR, Seo MJ, Youn YN, Park HJ, Yasunaga-Aoki C, Yu YM. Insecticidal effect of controlled release formulations of etofenprox based on nano-bio technique. Journal of the Faculty of Agriculture, Kyushu University, 2011; 56(1): 33–40.
51. Frandsen MVV, Pedersen MS, Zellweger M, Gouin S, Roorda SD, Phan TQC. 2011; U.S. Patent Application No. 13/057,773.
52. Anjali CH, Khan SS, Margulis-Goshen K, Magdassi S, Mukherjee A, Chandrasekaran N. Formulation of water-dispersible nanopermethrin for larvicidal applications. Ecotoxicology and Environmental Safety, 2010; 73(8): 1932–1936.
53. Isiklan N. Controlled release of insecticide carbaryl from crosslinked carboxymethylcellulose beads. Journal of Applied Polymer Science, 2006; 99(4): 1310–1319.
54. Medina-Pérez G, Fernández-Luqueño F, Campos-Montiel RG, Sánchez-López KB, Afanador-Barajas LN, Prince L. Nanotechnology in crop protection: status and future trends. Nano-Biopesticides Today and Future Perspectives, 2019; Jan 1: 17–45.
55. Hellmann C, Greiner A, Wendorff JH. Design of pheromone releasing nanofibers for plant protection. Polymers for Advanced Technologies, 2011; 22(4): 407–413.
56. Rudzinski WE, Chipuk T, Dave AM, Kumbar SG, Aminabhavi TM. pH-sensitive acrylic-based copolymeric hydrogels for the controlled release of a pesticide and a micronutrient. Journal of Applied Polymer Science, 2003; 87(3): 394–403.
57. Kok FN, Wilkins RM, Cain RB, Arica MY, Alaeddinoglu G, Hasirci VA. Controlled release of aldicarb from lignin loaded ionotropic hydrogel microspheres. Journal of Microencapsulation, 1999; 16(5): 613–623.
58. Fernández-Pérez J, Ahearne M. The impact of decellularization methods on extracellular matrix derived hydrogels. Scientific Reports, 2019; 9(1): 1–2.
59. Shakil NA, Singh MK, Pandey A, Kumar J, Pankaj, Parmar VS, Singh MK, Pandey RP, Watterson AC. Development of poly (ethylene glycol) based amphiphilic copolymers for controlled release delivery of carbofuran. Journal of Macromolecular Science, Part A: Pure and Applied Chemistry, 2010; 47(3): 241–247.

60. Liu Y, Laks P, Heiden P. Controlled release of biocides in solid wood. II. Efficacy against *Trametes versicolor* and *Gloeophyllum trabeum* wood decay fungi. Journal of Applied Polymer Science, 2002; 86(3): 608–614.

61. Wright JE. Formulation for insect sex pheromone dispersion. Patent number US. 1997; 5670145.

62. Quaglia F, Barbato F, De Rosa G, Granata E, Miro A, La Rotonda MI. Reduction of the environmental impact of pesticides: waxy microspheres encapsulating the insecticide carbaryl. Journal of Agricultural and Food Chemistry, 2001; 49(10): 4808–4812.

63. Chin CP, Wu HS, Wang SS. New approach to pesticide delivery using nanosuspensions: research and applications. Industrial & Engineering Chemistry Research, 2011; 50(12): 7637–7643.

64. Elek N, Hoffman R, Raviv U, Resh R, Ishaaya I, Magdassi S. Novaluron nanoparticles: formation and potential use in controlling agricultural insect pests. Colloids and Surfaces A: Physicochemical and Engineering Aspects, 2010; 372(1–3): 66–72.

65. Balaji AP, Ashu A, Manigandan S, Sastry TP, Mukherjee A, Chandrasekaran N. Polymeric nanoencapsulation of insect repellent: evaluation of its bioefficacy on *Culex quinquefasciatus* mosquito population and effective impregnation onto cotton fabrics for insect repellent clothing. Journal of King Saud University-Science, 2017; 29(4): 517–527.

66. Kang MA, Seo MJ, Hwang IC, Jang C, Park HJ, Yu YM, Youn YN. Insecticidal activity and feeding behavior of the green peach aphid, *Myzus persicae*, after treatment with nano types of pyrifluquinazon. Journal of Asia-Pacific Entomology, 2012; 15(4): 533–541.

67. Jamal M, Moharramipour S, Zandi M, Negahban M. Efficacy of nanoencapsulated formulation of essential oil from *Carum copticum* seeds on feeding behavior of *Plutella xylostella* (Lep.: Plutellidae). Journal of Entomological Society of Iran, 2013; 33(1): 23–31.

68. Yang FL, Li XG, Zhu F, Lei CL. Structural characterization of nanoparticles loaded with garlic essential oil and their insecticidal activity against *Tribolium castaneum* (Herbst) (Coleoptera: Tenebrionidae). Journal of Agricultural and Food Chemistry, 2009; 57(21): 10156–10162.

69. Feng BH, Peng LF. Synthesis and characterization of carboxymethyl chitosan carrying ricinoleic functions as an emulsifier for azadirachtin. Carbohydrate Polymers, 88(2): 576–582.

70. Borase HP, Patil CD, Salunkhe RB, Narkhede CP, Salunke BK, Patil SV. Phyto-synthesized silver nanoparticles: a potent mosquito biolarvicidal agent. Journal of Nanomedicine and Biotherapeutic Discovery, 2013; 3(1): 1–7.

71. Santhoshkumar T, Rahuman AA, Rajakumar G, Marimuthu S, Bagavan A, Jayaseelan C, Zahir AA, Elango G, Kamaraj C. Synthesis of silver nanoparticles using *Nelumbo nucifera* leaf extract and its larvicidal activity against malaria and filariasis vectors. Parasitology Research, 2011; 108(3): 693–702.

72. Paula HC, Rodrigues ML, Ribeiro WL, Stadler AS, Paula RC, Abreu FO. Protective effect of cashew gum nanoparticles on natural larvicide from *Moringa oleifera* seeds. Journal of Applied Polymer Science, 2012; 124(3): 1778–1784.

73. Lai F, Wissing SA, Müller RH, Fadda AM. *Artemisia arborescens* L essential oil-loaded solid lipid nanoparticles for potential agricultural application: preparation and characterization. Aaps Pharmscitech, 2006; 7(1): E10–E18.

74. Chitra G, Balasubramani G, Ramkumar R, Sowmiya R, Perumal P. *Mukia maderaspatana* (Cucurbitaceae) extract-mediated synthesis of silver nanoparticles to

control *Culex quinquefasciatus* and *Aedes aegypti* (Diptera: Culicidae). Parasitology Research, 2015; 114(4): 1407–1415.

75. Patil CD, Patil SV, Borase HP, Salunke BK, Salunkhe RB. Larvicidal activity of silver nanoparticles synthesized using *Plumeria rubra* plant latex against *Aedes aegypti* and *Anopheles stephensi*. Parasitology Research, 2012; 110(5): 1815–1822.

76. Chandra JH, Raj LA, Namasivayam SK, Bharani RA. Improved pesticidal activity of fungal metabolite from *Nomureae rileyi* with chitosan nanoparticles. In: International Conference on Advanced Nanomaterials & Emerging Engineering Technologies, Chennai, India, 2013; pp. 387–390. doi: 10.1109/ICANMEET.2013.6609326

77. Soni N, Prakash S. Efficacy of fungus mediated silver and gold nanoparticles against *Aedes aegypti* larvae. Parasitology Research, 2012; 110(1): 175–184.

78. Soni N, Prakash S. Factors affecting the geometry of silver nanoparticles synthesis in *Chrysosporium tropicum* and *Fusarium oxysporum*. American Journal of Nanotechnology, 2011; 2(1): 112–121.

79. Hadapad AB, Hire RS, Vijayalakshmi N, Dongre TK. Sustained-release biopolymer-based formulations for *Bacillus sphaericus* Neide ISPC-8. Journal of Pest Science, 2011; 84(2): 249–255.

80. Barik TK, Kamaraju R, Gowswami A. Silica nanoparticle: a potential new insecticide for mosquito vector control. Parasitology Research, 2012; 111(3): 1075–1083.

81. Goswami A, Roy I, Sengupta S, Debnath N. Novel applications of solid and liquid formulations of nanoparticles against insect pests and pathogens. Thin Solid Films, 2010; 519(3): 1252–1257.

82. Stadler T, Buteler M, Weaver DK. Novel use of nanostructured alumina as an insecticide. Pest Management Science: formerly Pesticide Science, 2010; 66(6): 577–579.

83. Rouhani M, Samih MA, Kalantari S. Insecticide effect of silver and zinc nanoparticles against *Aphis nerii* Boyer De Fonscolombe (Hemiptera: Aphididae). Chilean Journal of Agricultural Research, 2012; 72(4): 590.

84. Chakravarthy AK, Bhattacharyya A, Shashank PR, Epidi TT, Doddabasappa B, Mandal SK. DNA-tagged nano gold: a new tool for the control of the armyworm, *Spodoptera litura* Fab. (Lepidoptera: Noctuidae). African Journal of Biotechnology, 2012; 11(38): 9295–9301.

85. Rai M, Ingle A. Role of nanotechnology in agriculture with special reference to management of insect pests. Applied Microbiology and Biotechnology, 2012; 94(2): 287–293.

86. Kumar M, Shamsi TN, Parveen R, Fatima S. Application of nanotechnology in enhancement of crop productivity and integrated pest management. In: Nanotechnology. 2017; Springer, Singapore. pp. 361–371.

87. Elling AA. Major emerging problems with minor *Meloidogyne* species. Phytopathology, 2013; 103(11): 1092–1102.

88. Kumar V, Khan MR, Walia RK. Crop loss estimations due to plant-parasitic nematodes in major crops in India. National Academy of Science Letters, 2020; 43(5): 409–412.

89. Chakraborty S, Tiedemann AV, Teng PS. Climate change: potential impact on plant diseases. Environmental Pollution, 2000; 108(3): 317–326.

90. Decraemer W, Hunt DJ. Structure and classification. In: Plant Nematology R N Perry, M Moens (Eds.). 2006; CABI, Wallingford, UK. pp. 3–32.

91. Kyndt T, Zemene HY, Haeck A, Singh R, De Vleesschauwer D, Denil S, De Meyer T, Höfte M, Demeestere K, Gheysen G. Below-ground attack by the root knot nematode *Meloidogyne graminicola* predisposes rice to blast disease. Molecular Plant–Microbe Interactions, 2017; 30(3): 255–266.

92. Rocha LF, Pimentel MF, Bailey J, Wyciskalla T, Davidson D, Fakhoury AM, Bond JP. Impact of wheat on soybean cyst nematode population density in double-cropping soybean production. Frontiers in Plant Science, 2021; 12: 640714.

93. Sabry AK. Role of nanotechnology applications in plant-parasitic nematode control. In: Nanobiotechnology Applications in Plant Protection. 2019; Springer, Cham. pp. 223–240.

94. Bhau BS, Phukon P, Ahmed R, Gogoi B, Borah B, Baruah J, Sharma DK, Wann SB. A novel tool of nanotechnology: nanoparticle mediated control of nematode infection in plants. In: Microbial Inoculants in Sustainable Agricultural Productivity. 2016; Springer, New Delhi. pp. 253–269.

95. Chariou PL, Steinmetz NF. Delivery of pesticides to plant parasitic nematodes using tobacco mild green mosaic virus as a nanocarrier. ACS Nano, 2017; 11(5): 4719–4730.

96. Cheng C, Qin J, Wu C, Lei M, Wang Y, Zhang L. Suppressing a plant-parasitic nematode with frugivorous behavior by fungal transformation of a Bt cry gene. Microbial Cell Factories, 2018; 17(1): 1–4.

97. Fu Z, Chen K, Li L, Zhao F, Wang Y, Wang M, Shen Y, Cui H, Liu D, Guo X. Spherical and spindle-like abamectin-loaded nanoparticles by flash nanoprecipitation for southern root-knot nematode control: preparation and characterization. Nanomaterials, 2018; 8(6): 449.

98. Abdellatif KF, Abdelfattah RH, El-Ansary MS. Green nanoparticles engineering on root-knot nematode infecting eggplants and their effect on plant DNA modification. Iranian Journal of Biotechnology, 2016; 14(4): 250.

99. Nassar AM. Effectiveness of silver nano-particles of extracts of *Urtica urens* (Urticaceae) against root-knot nematode *Meloidogyne incognita*. Asian Journal of Nematology, 2016; 5: 14–19.

100. Abbassy MA, Abdel-Rasoul MA, Nassar AM, Soliman BS. Nematicidal activity of silver nanoparticles of botanical products against root-knot nematode, *Meloidogyne incognita*. Archives of Phytopathology and Plant Protection, 2017; 50(17–18): 909–926.

101. Memon AR, Schröder P. Implications of metal accumulation mechanisms to phytoremediation. Environmental Science and Pollution Research, 2009; 16(2): 162–175.

102. Lim D, Roh JY, Eom HJ, Choi JY, Hyun J, Choi J. Oxidative stress-related PMK-1 P38 MAPK activation as a mechanism for toxicity of silver nanoparticles to reproduction in the nematode *Caenorhabditis elegans*. Environmental Toxicology and Chemistry, 2012; 31(3): 585–592.

103. Wang H, Wick RL, Xing B. Toxicity of nanoparticulate and bulk ZnO, Al_2O_3 and TiO_2 to the nematode *Caenorhabditis elegans*. Environmental Pollution, 2009; 157(4): 1171–1177.

104. Wu S, Lu J, Rui Q, Yu S, Cai T, Wang D. Aluminum nanoparticle exposure in L1 larvae results in more severe lethality toxicity than in L4 larvae or young adults by strengthening the formation of stress response and intestinal lipofuscin accumulation in nematodes. Environmental Toxicology and Pharmacology, 2011; 31(1): 179–188.

105. Pluskota A, Horzowski E, Bossinger O, von Mikecz A. *Caenorhabditis elegans* nanoparticle-bio-interactions become transparent: silica-nanoparticles induce reproductive senescence. PLoS One, 2009; 4(8): e6622.

106. Roh JY, Sim SJ, Yi J, Park K, Chung KH, Ryu DY, Choi J. Ecotoxicity of silver nanoparticles on the soil nematode *Caenorhabditis elegans* using functional ecotoxicogenomics. Environmental Science & Technology, 2009; 43(10): 3933–3940.

107. Jo YK, Starr JL, Deng Y. Use of silver nanoparticles for nematode control on the Bermuda grass putting green. Turfgrass and Environmental Research Online, 2013; 12(2): 22–24.
108. Ardakani AS. Toxicity of silver, titanium and silicon nanoparticles on the root-knot nematode, *Meloidogyne incognita* and growth parameters of tomato. Nematology, 2013; 15(6): 671–677.
109. El-Deen AN, El-Deeb BA. Effectiveness of silver nanoparticles against root-knot nematode, *Meloidogyne incognita* infecting tomato under greenhouse conditions. Journal of Agricultural Science, 2018; 10: 148–156.
110. Thakur RK, Shirkot P. Potential of biogold nanoparticles to control plant pathogenic nematodes. Journal of Bioanalysis and Biomedicine, 2017; 9(4): 220–222.
111. Ahamed M, Posgai R, Gorey TJ, Nielsen M, Hussain SM, Rowe JJ. Silver nanoparticles induced heat shock protein 70, oxidative stress and apoptosis in *Drosophila melanogaster*. Toxicology and Applied Pharmacology, 2010; 242(3): 263–269.
112. Dalakouras A, Wassenegger M, Dadami E, Ganopoulos I, Pappas ML, Papadopoulou K. Genetically modified organism-free RNA interference: exogenous application of RNA molecules in plants. Plant Physiology, 2020; 182(1): 38–50.
113. Jain RG, Fletcher SJ, Manzie N, Robinson KE, Li P, Lu E, Brosnan CA, Xu ZP, Mitter N. Foliar application of clay-delivered RNA interference for whitefly control. Nature Plants, 2022; 8(5): 535–548.
114. Cao J, Guenther RH, Sit TL, Lommel SA, Opperman CH, Willoughby JA. Development of abamectin loaded plant virus nanoparticles for efficacious plant parasitic nematode control. ACS Applied Materials and Interfaces, 2015; 7(18): 9546–9553.

Index

For Product Safety Concerns and Information please contact our EU
representative GPSR@taylorandfrancis.com
Taylor & Francis Verlag GmbH, Kaufingerstraße 24, 80331 München, Germany